WORLD of MACHINES

AN INTRODUCTION TO SIMPLE MACHINES
for Young Scientists

WORLD of MACHINES

AN INTRODUCTION TO SIMPLE MACHINES
for Young Scientists

HERON
BOOKS

Published by
Heron Books, Inc.
20950 SW Rock Creek Road
Sheridan, OR 97378

heronbooks.com

———————————

Special thanks to all the teachers and students who
provided feedback instrumental to this edition.

———————————

Printed in the USA

18 July 2022

At Heron Books, we think learning should be engaging and fun. It should be hands-on and allow students to move at their own pace.

To facilitate this we have created a learning guide that will help any student progress through this book, chapter by chapter, with confidence and interest.

Get learning guides at
heronbooks.com/learningguides.

For teacher resources,
such as a final exam, email
teacherresources@heronbooks.com.

We would love to hear from you!
Email us at *feedback@heronbooks.com.*

Your YOUNG SCIENTIST JOURNAL

Scientists love to explore the world and how things in it work. They like to go new places and discover things they've never seen before.

They also like to keep track of what they find. They often fill books with notes and drawings of what they see, and include their thoughts and questions about it. These books are called *science journals*.

What's fun about a science journal is that you can use it to draw pictures or sketches of things that interest you. You can write down ideas you have about things, make maps, write down questions you have and things you want to find out more about. You might even stick in it samples of things you find—flowers, bugs, leaves, feathers, spider's webs—who knows what?

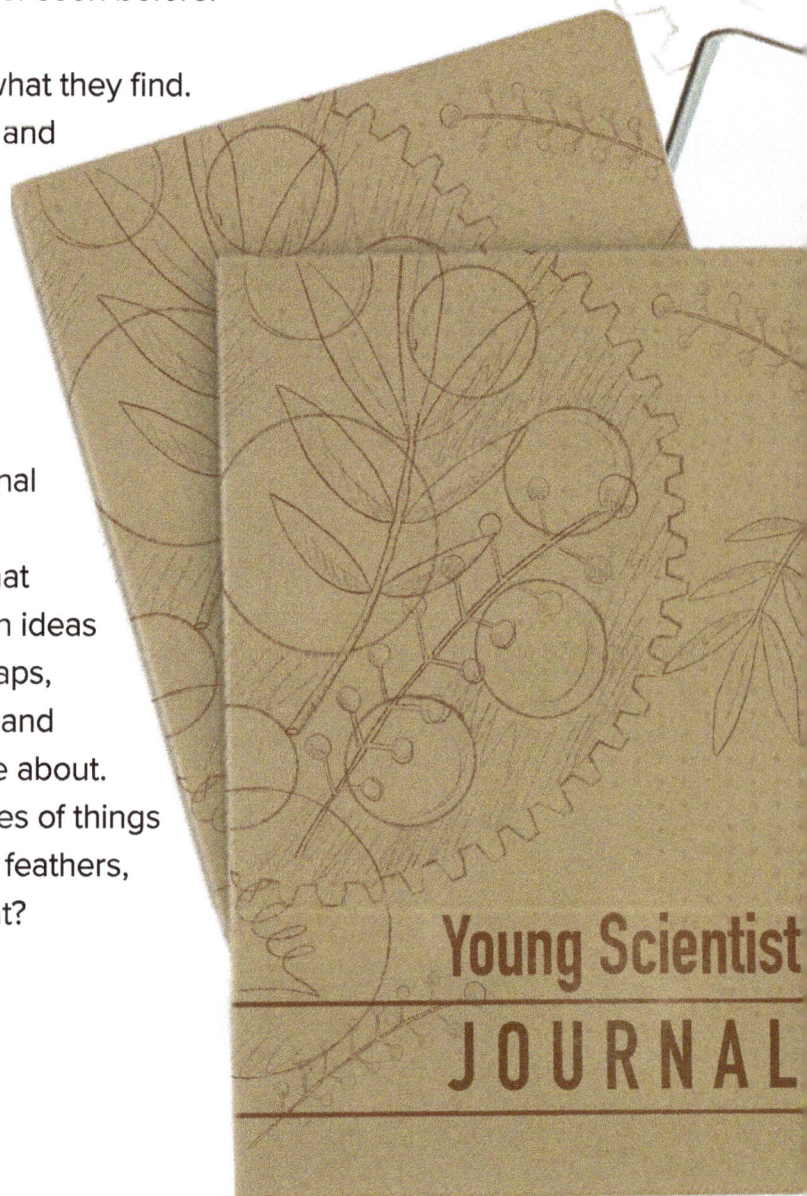

Young Scientist

JOURNAL

The learning guide that goes with this book will sometimes ask you to look at things and make notes or drawings in a journal of your own.

Whatever you put in your science journal, it will be full of your own personal discoveries. No two journals are alike.

You can use a journal like the one shown here, or you can use a notebook of your choice. You might even want to make your own science journal and use that.

Whichever type of journal you choose, it will be a place to keep drawings and notes about what you are finding out about the world and how it works.

So get ahold of a science journal, or make one, and then get going to see what you can find out. Who knows what might be waiting for you?

IN THIS BOOK

1 MACHINES ARE ALL AROUND US   1

What Are the Six Simple Machines?   4

2 MATTER AND MASS   10

Matter   12

Mass   13

Let's Do This: How Much Mass?   16

3 DENSITY   18

Comparing Densities   20

Why Different Densities?   22

4 INERTIA   24

Inertia and Mass   26

Let's Do This: Mass and Inertia   29

5 FORCE   30

Measuring Force   33

Let's Do This: Force and Inertia   36

Let's Do This: Measuring Force   37

6 FORCE AND MOTION   38

Newton's First Law   40

Newton's Second Law   42

Newton's Third Law   44

7 GRAVITY 48
Gravity, Weight and Mass 52

8 FRICTION 54
What Causes Friction? 56
Reducing Friction 57
Increasing Friction 58
How Friction Affects Us 59
Let's Do This: Friction and Surfaces 62

9 MACHINES & WORK 64
What Is Work? 66
Work Is Not the Same as Force 66

10 LEVERS 68
Mechanical Advantage 69
Fulcrum 72
Let's Do This: Using a Lever 73

11 THREE KINDS OF LEVERS 74
First Class Lever 74
Second Class Lever 75
Third Class Lever 75
Let's Do This: First Class Lever 76
Let's Do This: Second Class Lever 78
Let's Do This: Third Class Lever 80

12 WHEEL AND AXLE 82

Winches 84

Bicycles 84

Other Wheel and Axle Examples 85

Surprise! 85

Let's Do This: Wheel and Axle 86

13 PULLEY 88

Fixed Pulleys 88

Moveable Pulleys 89

Let's Do This: Pulley 92

Let's Do This: Block and Tackle 94

14 INCLINED PLANE 96

Let's Do This: Ramp 98

15 WEDGE 100

Let's Do This: Wedge 103

16 SCREW 104

Let's Do This: Screw 106

17 COMBINING SIMPLE MACHINES 108

Let's Do This: Make a Compound Machine 111

Teacher Tips 112

1 CHAPTER MACHINES ARE ALL

AROUND US

All around the world, people do interesting jobs every day.

They build houses, bridges and skyscrapers.

They plow fields to grow wheat and corn.

They move things from one place to another.

They cook, teach, fly jets and make music.

Machines are all around us. In almost every job that people do, they use machines to help them do their work.

A **machine** is a piece of equipment used to make work easier. Machines allow people to get work done with less effort. They also make it possible to do jobs that are difficult or complicated.

Imagine trying to cut the grass on a playing field without a lawnmower!

Some of the machines people use the most are referred to as "simple machines." Humans have been using these since ancient times, and they are still in use every day all around the world.

Simple here means "not complicated, having only a few parts."

A car or a computer is a machine made up of many, many parts and is *complicated*. A **simple machine** has just one or two parts.

There are only six simple machines. Although they have few parts, they are helpful in thousands of ways because they reduce the amount of effort needed to do something. A simple machine can make it possible to carry out a task that would be much harder without it.

Almost all the more complicated machines we use, from electric fans to tractors and cranes, are some combination of these six simple machines working together.

If you like designing and building things, you may already know something about these simple machines. If you haven't spent much time designing and building, learning about them may help you get started.

An engineer designs and helps build things like roads, bridges, dams, tunnels, airports and even spacecraft. Engineers also design and build tools and other machines.

WHAT ARE THE SIX SIMPLE MACHINES?

Lever

First let's talk about a lever. A **lever** is used to lift things.

A lever has two parts.

1. A bar or beam, or something similar that doesn't bend easily.

2. A support the bar or beam sits on.

When the bar is pushed down, this simple machine can be used to lift things with much less effort.

We often see levers on playgrounds. Remember playing on a seesaw or teeter-totter? With this kind of lever, a child can lift another off the ground with little effort.

It might be difficult to lift a big boulder without using a lever.

Even something as simple as a bottle opener is a lever.

A lever helps you lift things more easily, using less effort.

Wheel and Axle

If you've ever ridden a tricycle, you are familiar with the next simple machine, a wheel and axle. It also has two parts.

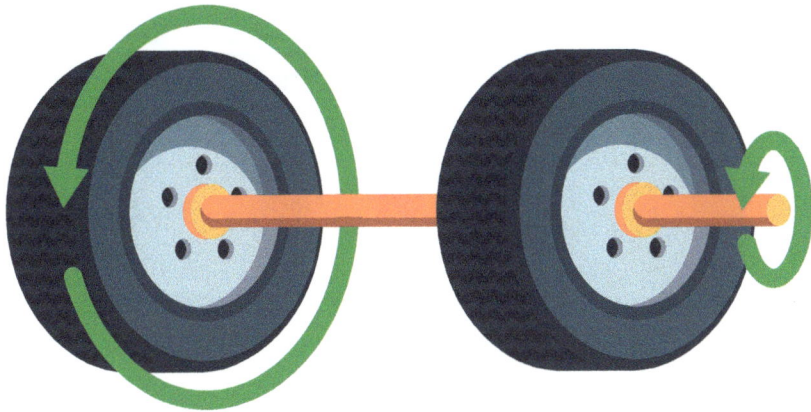

Any wheel-shaped, circular object with a round rod through its center is a **wheel and axle**.

A wheel and axle can be used to move things. Look at a tricycle, for example. Pedaling turns an axle. The wheel, being attached to the axle, then also turns, and this moves the tricycle along.

A ferris wheel is another great example of a wheel and axle. A motor turns a rod through the center, and this makes the big, circular wheel go around and around.

Pulley

If you've ever raised or lowered the window blinds in your home, you may well have been using our next simple machine. A **pulley** is a wheel with a groove in it. Using a rope or cord with the pulley makes it possible to lift a heavy load more easily.

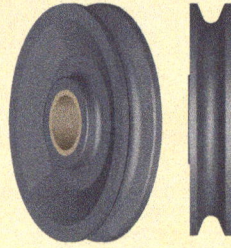

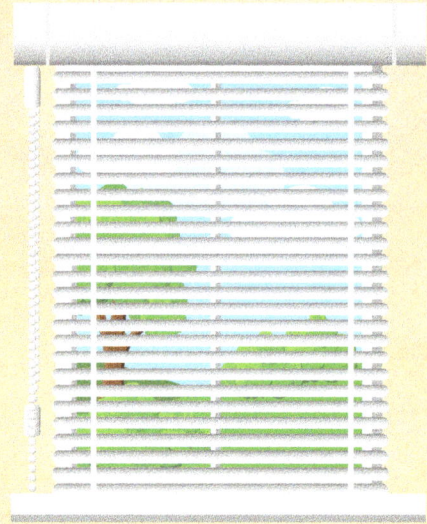

Like the lever, and the wheel and axle, the pulley can be used to help people move things.

Sometimes the branch of a tree or even a metal hook can be used as a pulley.

In addition to lifting heavy things, a pulley and rope can be helpful in moving something to a high spot that is difficult to reach.

CLUBHOUSE

Inclined Plane

An **inclined plane** is a simple machine with only one part. A **plane** is a flat surface. **Inclined** means "slanted." So an inclined plane is a flat surface set at a slant. One end is higher than the other.

An inclined plane is used to connect a higher place to a lower one. It takes some of the weight of an object being moved up or down it, so it doesn't take as much effort to move something heavy.

All ramps are inclined planes. Some examples are loading ramps, wheelchair ramps and skateboard ramps.

If you think about it, a playground slide is also an inclined plane. It connects the top of the slide with the ground and allows a person to move easily from the top to the bottom.

Just like a lever, a wheel and axle, and a pulley, an inclined plane is a simple machine that makes moving things a lot easier.

Wedge

A **wedge** is an object that is thick at one end and thinner at the other. One of its surfaces is an inclined plane. A wedge looks like this.

Wedges are often used to push things apart. An example is the head of an axe, which can be used to chop down a tree or separate a log into smaller pieces for a fire.

The nose of an airplane is a wedge that helps the plane move more easily through the air.

The bow of a ship is a wedge that cuts through the water.

If you look carefully at the blade of a knife or the blades of a pair of scissors, you'll see that they are also wedges.

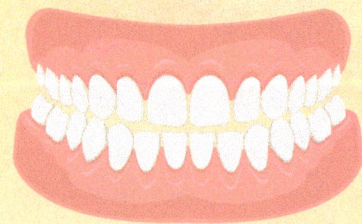

If you think about it, even your front teeth are wedges! They help you separate food into bites.

Screw

The last of the six simple machines is the **screw**, a cylinder with ridges spiraling around it.

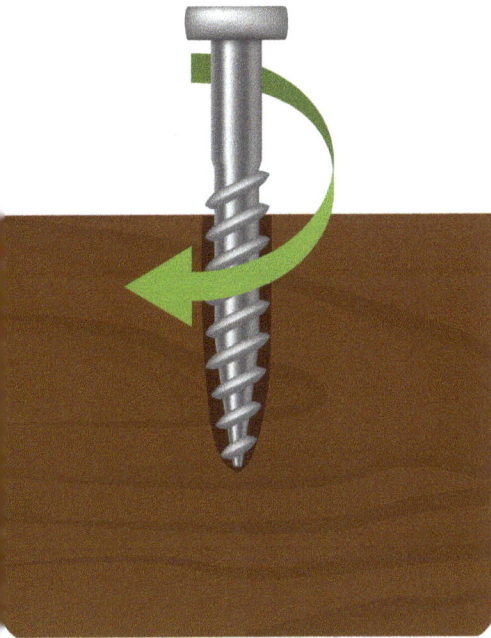

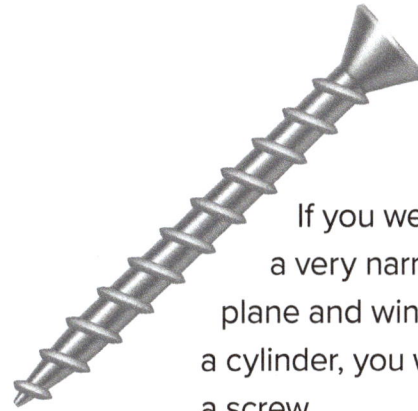

If you were to take a very narrow inclined plane and wind it around a cylinder, you would have a screw.

When a screw is turned, the turning force and the spiraling ridges move the screw into something.

A screw can be used to hold things tightly together. Screws will hold a bookshelf on a wall, for example.

The top of a jar is a type of screw. It helps you put a lid on tightly. The same is true of many bottles. Take a look at the top of a water bottle and you'll find a screw.

If you look closely at the base of a light bulb, you'll see it is a screw. It helps hold the bulb securely in the socket so electricity can flow into it safely.

Every day we use these six simple machines, the lever, wheel and axle, pulley, inclined plane, wedge, and screw to make work easier in many ways.

CHAPTER 2

MATTER AND MASS

Now that you have a pretty good idea of what the simple machines are, what else is there to know about them?

Well, we could try to understand *how* they work. For example, what happens when you use a ramp to move something? Why does the ramp make it possible to move a heavy load so much more easily?

What does a lever actually *do?* How does it change a force to make a job easier?

Understanding how and why these simple machines work gives us information we can use to be in better control of the world around us.

All your life you're going to be *doing* things.

Putting together a skateboard ramp.

You might be building a treehouse.

Or cleaning up your yard.

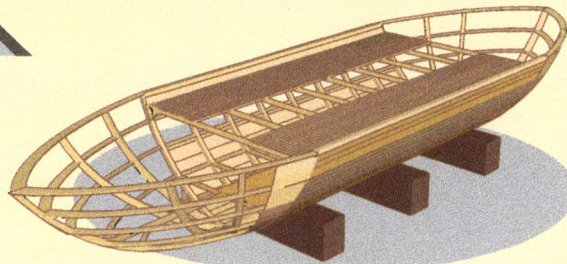

Building a raft or a boat.

Installing shelves to hold all the stuff in your room.

Deciding where to put a bike trail and how to design it to be challenging and fun.

You might even want to build your own house someday.

If you understand something about the physical world and how it works, you'll be able to do more things. Not only that, you'll have a lot more information you can use to turn all the great ideas you have into real things!

So, let's take a look at some information that might be useful about the physical world and how it works.

MATTER

The **physical world** is the world of trees, rocks, clouds, wind and water.

It's the world of cars, buildings, bridges and roads. It's the world of all the things we can see and touch.

Everything in the physical world is made of **matter.** Scientists use this word for everything we can see or touch.

All matter takes up space.

A book, for example, takes up space on the table.

Water takes up space in a glass.

Air takes up space in a balloon.

Paper, water and air are all matter.

MASS

Here's another word scientists use. **Mass** is the *amount of* matter in something.

While the words "matter" and "mass" are similar, they stand for different things.

When we look at a tree, we know it is matter because we can see it. It is part of the physical world.

But *how much* matter? A large tree would have a lot of mass. A small tree would have less mass.

The mass of something is the specific amount of matter it contains.

A piece of rock the size of your fist has a certain mass.

The same kind of rock that's twice that size has double that amount of matter, or double the mass.

On the other hand, a small chest of gold coins has more mass than a large bag of dollar bills. Why?

The amount of matter in something isn't just based on how big it is. It also depends on what kind of matter the object is made of.

For example, think about a balloon full of air compared to a bowling ball the same size.

Which do you think has more mass?

If you guessed the bowling ball, you're right! Seems kind of obvious, doesn't it?

But how could you tell ?

The bowling ball is harder to move around, it takes more of your effort to move it.

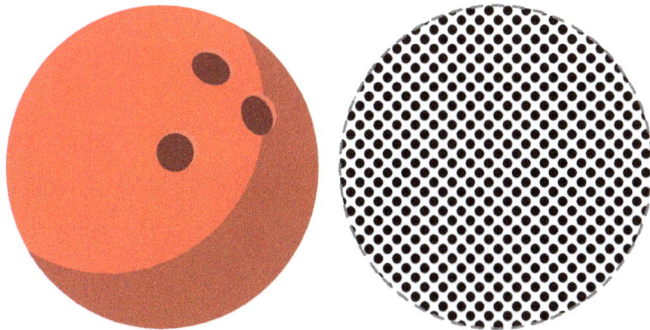

Right away this tells you that there's a lot more matter packed into that bowling ball than there is in the air-filled balloon.

It tells you that it has more mass.

HOW MUCH MASS?

For this activity you will need

- a brick

- a block of wood about the same size as the brick

- your science journal

Steps

1 Experiment with the brick and the wood block to answer these questions:

- Which is harder to pick up or push around?

- Which would be harder to throw or catch?

- Which of these objects has more mass?

2 In your journal, write what you found out.

3 Find three objects that are made of the same type of matter but are different in size.

4 Compare them and decide which has the most mass and which has the least.

5 In your journal, write what you found out.

6 Find three new objects that are about the same size but made of different types of matter.

7 Hold each object in your hand and swing it back and forth, then stop it from swinging. Decide which has the most mass and which has the least.

8 In your journal, write what you found out.

9 What have you noticed about what mass and matter have to do with one another? Write your thoughts in your journal.

3 DENSITY

Now we're going to talk about something else scientists have noticed about the objects in the physical world.

When an object is heavy for its size, it is said to be **dense**.

A golf ball is more dense than a ping pong ball. Even though they are the same size, the golf ball has more matter in it.

Density is the amount of matter in something compared to its size. High density means an object has a lot of matter squished together for the size. Low density means not much matter for the size.

As an example, a block of steel is very dense while the same amount of air is not.

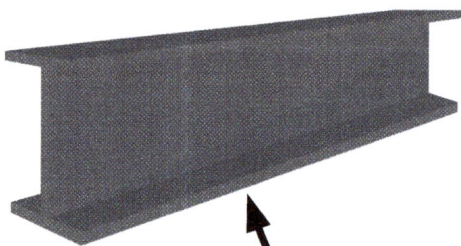

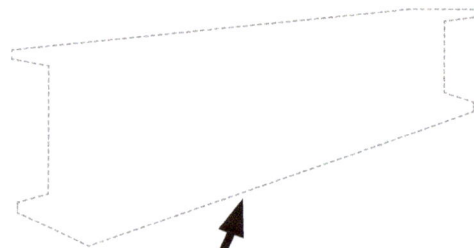

steel has a high density and air has a low density

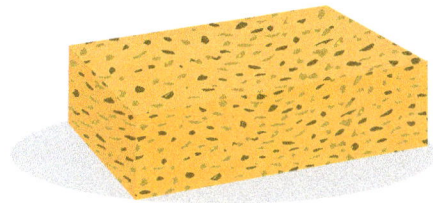

A brick is more dense than a sponge of the same size.

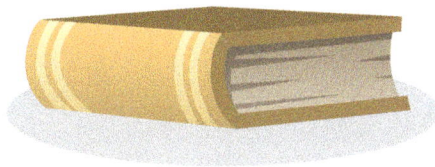

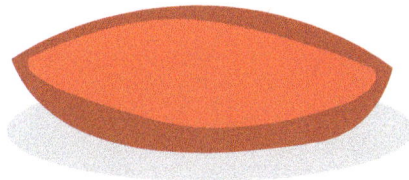

A book is more dense than a pillow of the same size.

COMPARING DENSITIES

A useful thing to know about density is that lighter things, things with lower density, are more likely to float.

Heavier things with higher densities are more likely to sink.

This is why a rubber duck will float while a rock the same size will not.

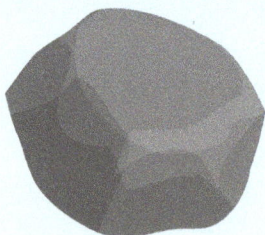

So how can you tell if one substance is more or less dense than another?

Suppose you have a cup of water and a cup of sand, and you want to compare them to see which is more dense, or has higher density.

First, make sure you have exactly the same amount of each. In this case, it's one cup. So they're the same size.

Then experiment to see which feels like it has more mass. You might try picking them up to see how heavy each one is.

A more scientific way would be to use a spring scale that measures mass in grams or kilograms.

The higher the density, the heavier something feels. It feels heavier because it has more mass.

WHY DIFFERENT DENSITIES?

Why is some matter more dense?

To answer that, let's talk about what matter is made of. Most likely you already know this. Matter is made up of **atoms,** pieces of matter so tiny that there are millions of them in the period at the end of this sentence.

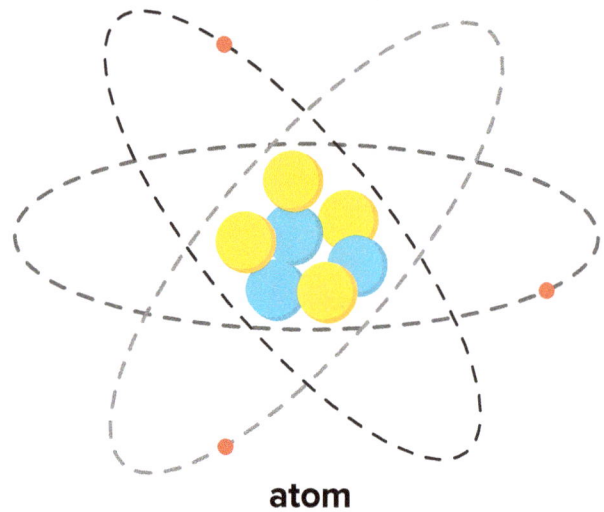

atom

Imagine you have a marshmallow and some clay, and you create a one-inch cylinder of each.

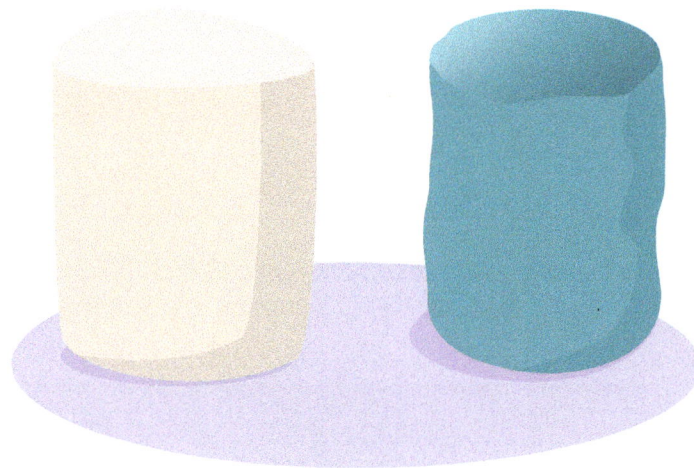

When you test them, you'll find that the clay is denser than the marshmallow.

Though they are the same size, the clay has more mass.

But *why* does it have more mass?

The difference comes from the fact that the atoms in the marshmallow have more space between them than the atoms in the clay.

This means there are far fewer atoms in the marshmallow cylinder.

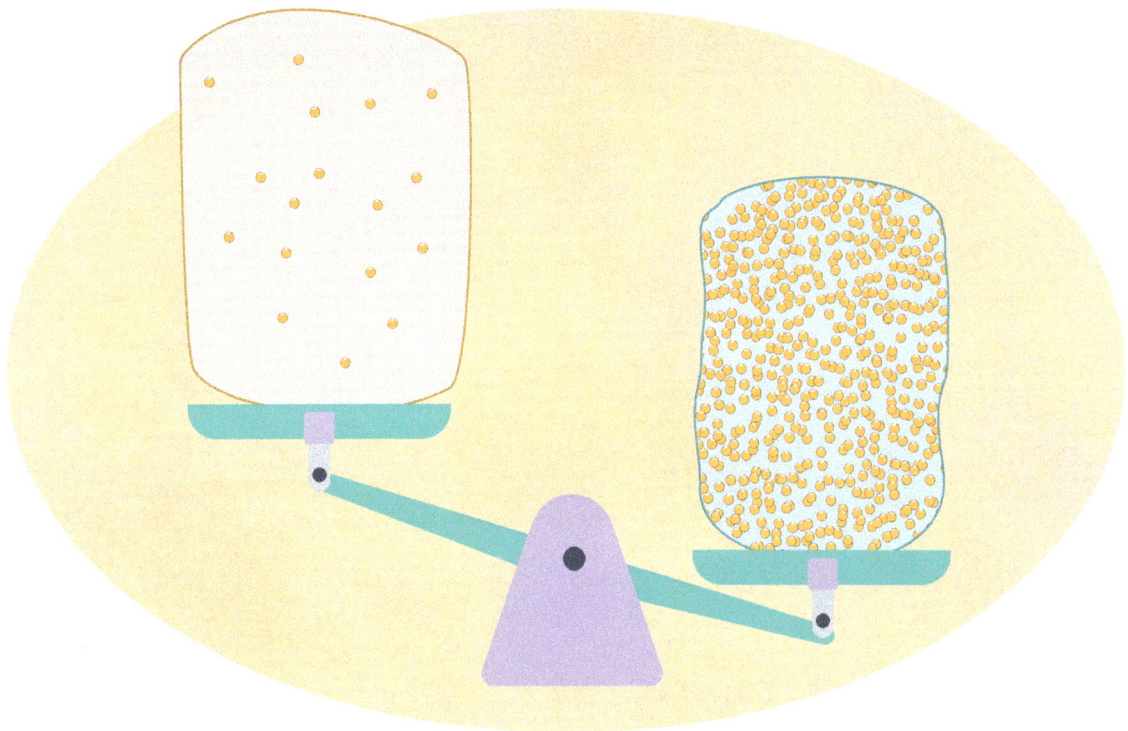

There are many more in the cylinder made of clay.

Density measures how close together the atoms in a piece of matter are. Are they all squished together or are they spread out with more space between them?

INERTIA

You've probably noticed that when an object is moving, it tends to keep moving unless something stops it. If you roll a bowling ball down the lane, it will keep on rolling and rolling until it knocks over the pins and hits the back wall behind them.

Objects that are not moving "want" to stay still.

Objects that are in motion "want" to keep moving.

You've probably also noticed that when something is *not* moving, it tends to stay still until something moves it. This happens all the time. Put your food on the table and watch it. Until you pick it up to eat it, it just sits there.

The scientific word for this is *inertia*. It is something that's true about all matter. When an object is at rest (not moving), it will tend to stay at rest. When it is moving, it will tend to keep moving.

Imagine someone throws you a baseball. When you catch it, you can feel that it "wants" to keep going. It pushes against your hand.

This piece of matter, the baseball, has inertia— once it is moving, it tends to keep moving.

Now imagine you want to ride a bicycle. It's parked. Unless you get it moving, it "wants" to stay where it is.

INERTIA AND MASS

The amount of inertia an object has depends on its mass. The more mass an object has, the more inertia it has, and the harder it is to get it moving or stop it.

If you've ever been bowling, you've seen this in action. When a bowling ball hits the bowling pins, they go flying. They don't stop the ball. Why?

When you push on the pedals to get it moving, the first push is the hardest because inertia holds the bike back from moving.

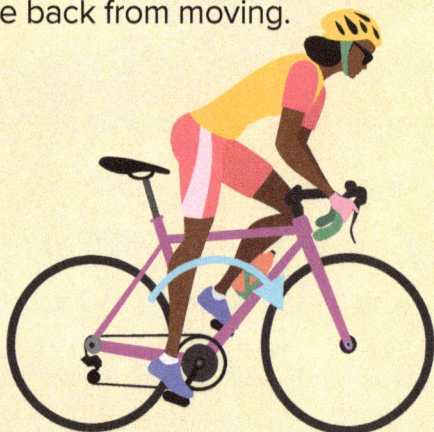

Once you get it moving, the bike wants to keep moving and pedaling gets easier.

The ball has more mass than the pins, so it has more inertia. Plus, it's already moving, which means it will tend to keep moving rather than be stopped by the pins. Its greater mass and the fact that it's already moving give it more inertia.

If you threw a wad of paper at a window, it would just bounce off. But if you threw a rock at that same window, there would probably be a different result.

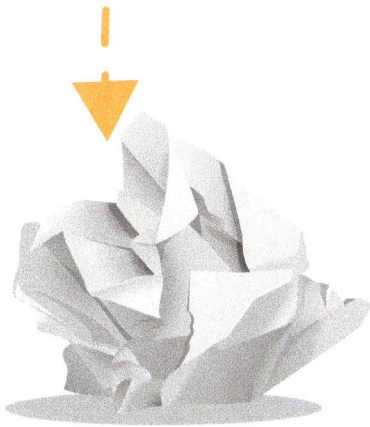

The wad of paper wants to keep moving because of its inertia. The glass wants to stay still because of its inertia. Who wins? The one with more mass. The window.

What about the rock? It wants to keep moving because of its inertia. The window wants to stay where it is because of its inertia. Who wins this time? The rock has more mass than the window, so it is likely to win the battle with the glass.

MASS AND INERTIA

For this activity you will need

- a brick

- several other moveable objects of different weights

- your science journal

Steps

1 Swing a brick in your hand and then try to stop the motion.

2 Hold the brick still and then try to move it suddenly.

3 Repeat these actions with lighter and heavier objects.

4 In your journal, make a list of the objects you tested. Put them in order from most mass and inertia to least mass and inertia.

5 CHAPTER FORCE

To overcome the inertia of an object and get it moving, or to stop it, there needs to be some kind of a push or a pull. In science, a push or a pull, no matter how weak or strong, is called a **force.**

A force is simply a push or a pull, whether it makes something move or not.

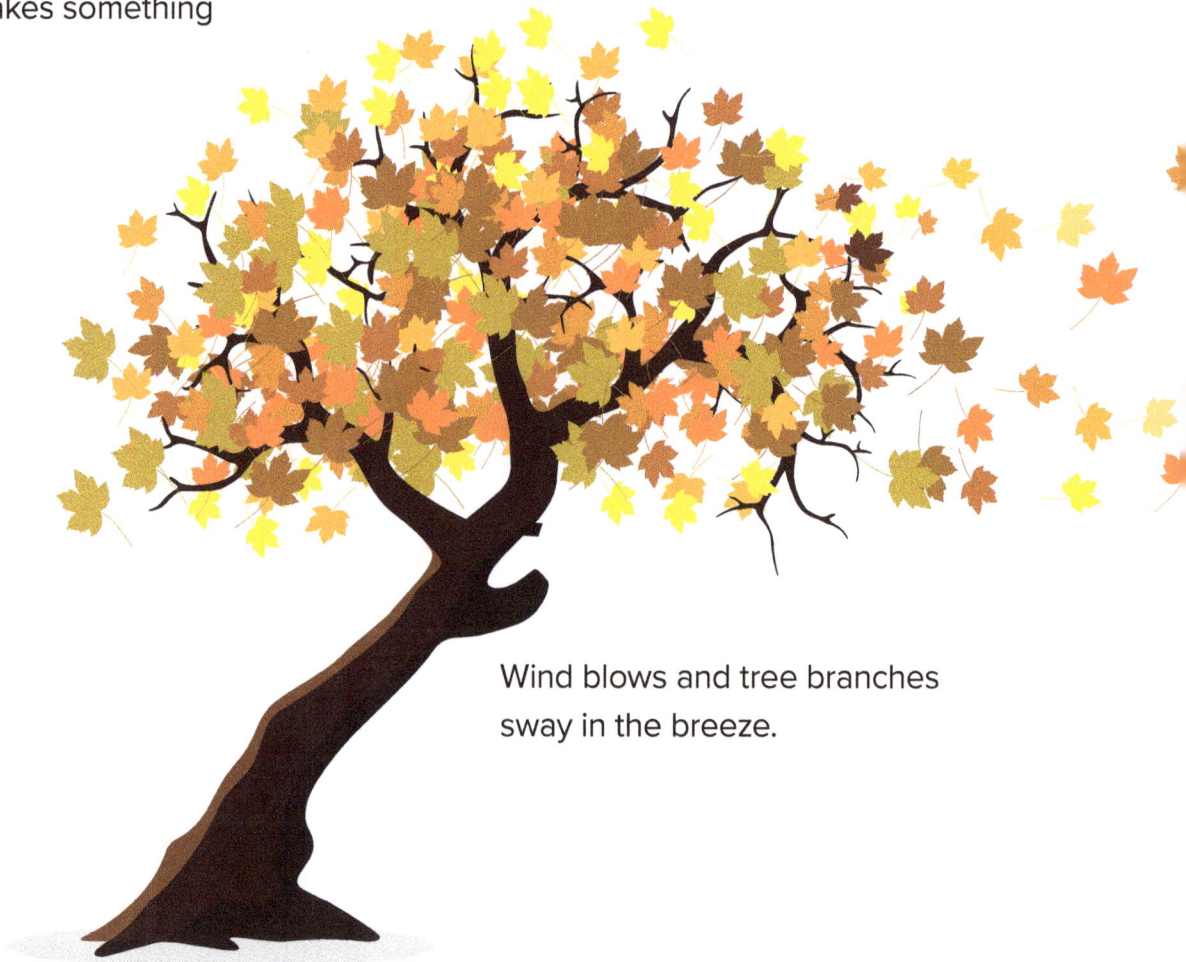

Wind blows and tree branches sway in the breeze.

The thing a force is pushing or pulling on may or may not move. For example, if you push on a wall, you are using force, but it probably won't be enough to move the wall.

When you push a friend on a swing, it moves away from you.

When you pull the brake lever on the handlebars, your bicycle slows down.

When you pull on your dog's leash, she comes in your direction.

Forces are at work all around us.

When you kick a ball, your foot applies a force to the ball and gets the ball moving.

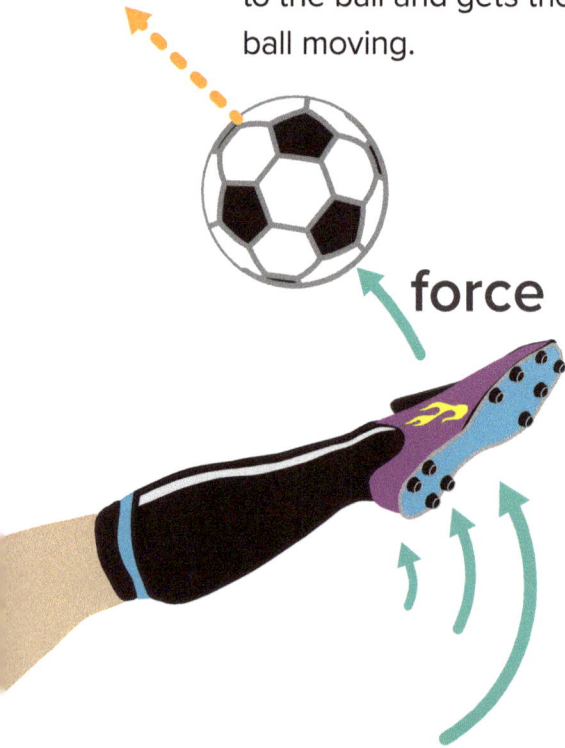

force

When you start to run, your feet apply force to the ground and you move.

force

A force goes in a direction. When you push something, it moves in the direction you push it.

If you push straight ahead, the object will move straight ahead, not to the right or left.

If you push an object straight ahead and (as sometimes happens) it doesn't move exactly straight ahead, it's because there is some other force acting to move it in a different direction.

Suppose you throw a ball in one direction, but a strong wind is blowing across that path.

The ball will get pushed by your throw and the wind at the same time.

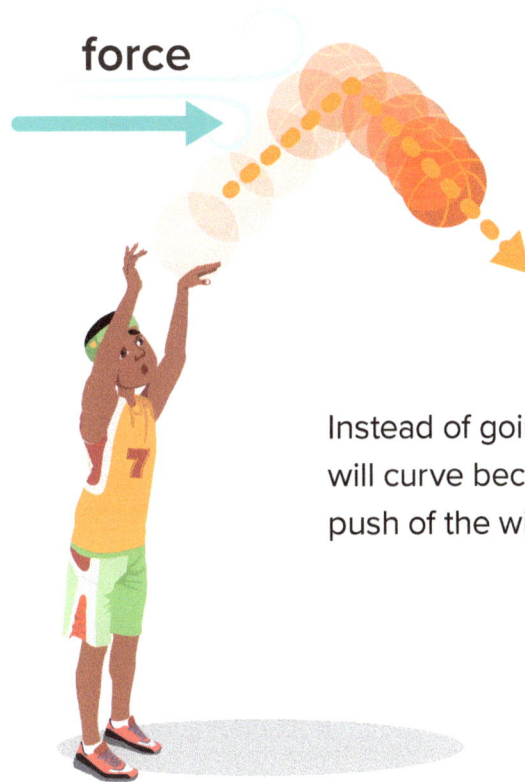

force

Instead of going straight, it will curve because of the push of the wind.

MEASURING FORCE

The more mass an object has, the more force is needed to move or stop it.

It's easier, for example, to push a small car than it is to push a truck because the small car has less mass. So, it takes less force to make it move.

Force is measured in units called **newtons.** (The abbreviation for newton is N). Newtons are named for a famous scientist named Isaac Newton.

One newton is the amount of force it takes to lift about 100 grams.

A lemon weighs about 100 grams, so the amount of force needed to pick up that lemon is 1 newton.

You can measure force on a spring scale.

If a cell phone weighed 200 grams (less than half a pound), it would take about 2 N to lift it.

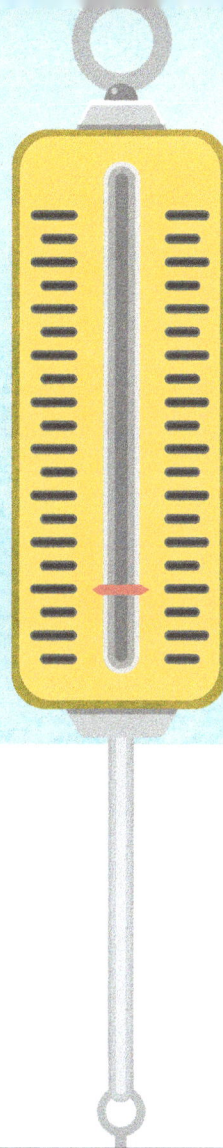

How many newtons would it take to lift a super-light, 1,000 gram laptop? That's right, 10 newtons!

FORCE AND INERTIA

For this activity you will need

- 5 different objects of different sizes, including at least one that's unmovable, such as a wall

- your science journal

Steps

1 Collect at least five objects of different sizes to use to demonstrate force. Include at least one thing that you can't move.

2 For each object, use force to try to overcome inertia and change its motion. Notice if you are pushing or pulling.

3 In your journal, write down each object, what force you used (pushing or pulling), and what happened.

MEASURING FORCE

Let's Do This!

For this activity you will need

- spring scale that measures both mass (up to 1,000 grams) and force (up to 10 newtons)

- 500 gram weight

- 1,000 gram weight

Steps

1 Look at the spring scale and see how it works.

2 Using the spring scale, see how hard you need to pull to get

- 1 newton

- 3 newtons

- 5 newtons

- 10 newtons

3 See how many newtons it takes to lift

- 500 grams

- 1500 grams

4 Decide how many newtons it would take to lift

- 200 grams

- 1000 grams

Then try it and see if you were right!

6 CHAPTER FORCE AND MOTION

We have seen that objects move in predictable ways. If you kick a ball on a flat surface, it will go in the direction you kicked it, and it will keep going until something stops it. If the ball stops moving, it will just sit there until something moves it.

Isaac Newton, who lived during the 1600s, was one of the world's greatest scientists. He discovered gravity, figured out the shape of the earth, and was the first to find out how fast sound traveled. Newton was a keen observer. He discovered three natural laws that explain almost everything about the way physical objects move.

Natural law: water
freezes at 32 degrees.

What is a **natural law**? It's something that happens in nature—the physical world. It is something that has been observed to happen over and over by many people, the same way every time. In fact, it's been observed to happen the same way so often that we say it is always true. This is why we can call it a law.

Natural laws are also called **scientific laws**. We know them because scientists have done a lot of observation and experimenting to find out how the physical world works. They've done a lot of work finding out what are the things about the physical world that always happen the same way.

The natural (or scientific) laws that Newton discovered tell us how force and motion always work. We call them **Newton's Laws of Motion**.

NEWTON'S FIRST LAW

Newton's first law of motion has to do with inertia.

As we discussed earlier, all matter either keeps moving, or it stays still. This is called the **Law of Inertia**.

It has been observed over and over that any object that is moving tends to keep moving in the same direction.

Newton's First Law: An object in motion will continue to move in the same direction and at the same speed unless a force affects it. An object at rest will stay at rest until a force makes it move.

So what does it take to make a still object move?

What does it take to stop or change the direction of an object that's already in motion?

Any object that's at rest tends to stay that way.

You can always count on this happening.

It takes another force to act on it. In other words, you have to kick a soccer ball or push the pedals of your bike. To get the bike to stop, you also need to apply force. You need to use the brakes.

NEWTON'S SECOND LAW

Isaac Newton also observed some interesting things about how force and mass affect the motion of an object.

First of all, the more force, the more motion it causes. When you ride a bike, you use your legs to apply force to get it to move. The harder you push on the pedals, the faster it moves.

Even though we barely think about this because we're so used to it, this law helps us know how much force to use.

This brings us to the other part of Newton's Second Law, which tells us something else about how much force is needed. It tells us how mass and force work together.

If you pushed a basketball and a bowling ball with the same amount of force, the motion of the basketball would change a lot more than the motion of the bowling ball.

Newton's Second Law: A force on an object will cause it to move. Or, if it's already moving, will cause it to speed up, slow down or change direction.

The same idea works when you hit a baseball with a bat. The harder you hit the ball the farther it will go.

If you want to hit a home run, tapping the ball won't do the trick. You have to give it a good whack!

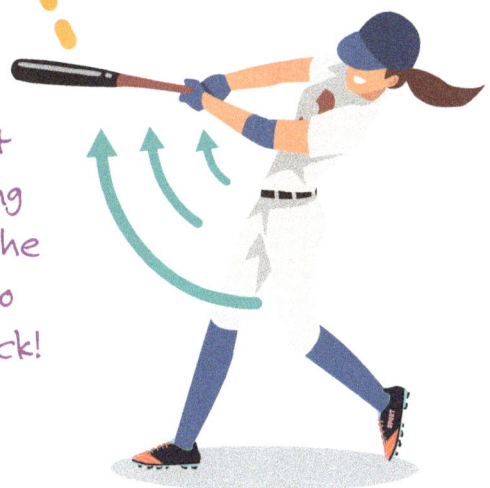

If both balls were rolling, which would take more force to stop?

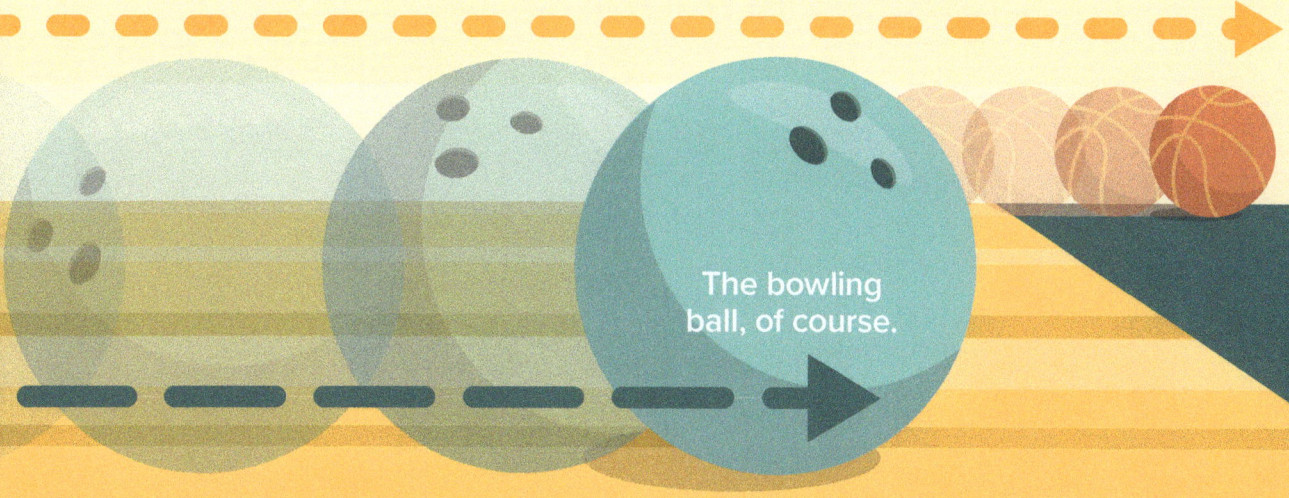

The bowling ball, of course.

The greater the force on the object, the more it will speed up, slow down or change direction. The lesser the force, the less it will speed up, slow down or change direction.

NEWTON'S THIRD LAW

This law tells us something very interesting about how the universe works, something you may or may not have noticed. Newton observed that forces always come in pairs. Every force goes in a direction, and there is an equal force in the opposite direction. They balance each other. He said it like this:

For every action there is always

an equal and opposite reaction.

In this law, "action" means a force and "reaction" means another opposite force that goes with the first one. Here are some examples.

reaction

action

You hit a baseball with a bat, and the bat pushes the ball into the air. The bat applies force to the ball.

reaction

At the same time the ball applies a force to the bat and slows it down.

action

The force on the bat is equal to the force on the ball and in the opposite direction. The bat doesn't slow down as much as the ball speeds up because the bat has more mass.

action

You blow up a balloon and let it go. The air coming out of the balloon is a force in one direction.

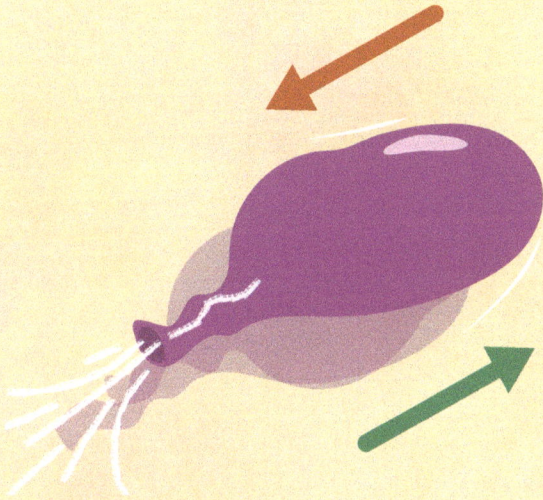

reaction

The balloon goes flying in the opposite direction. An equal and opposite action.

action

You stub your toe on a rock while you are running. Your toe applies a good-sized force, a push, to the rock.

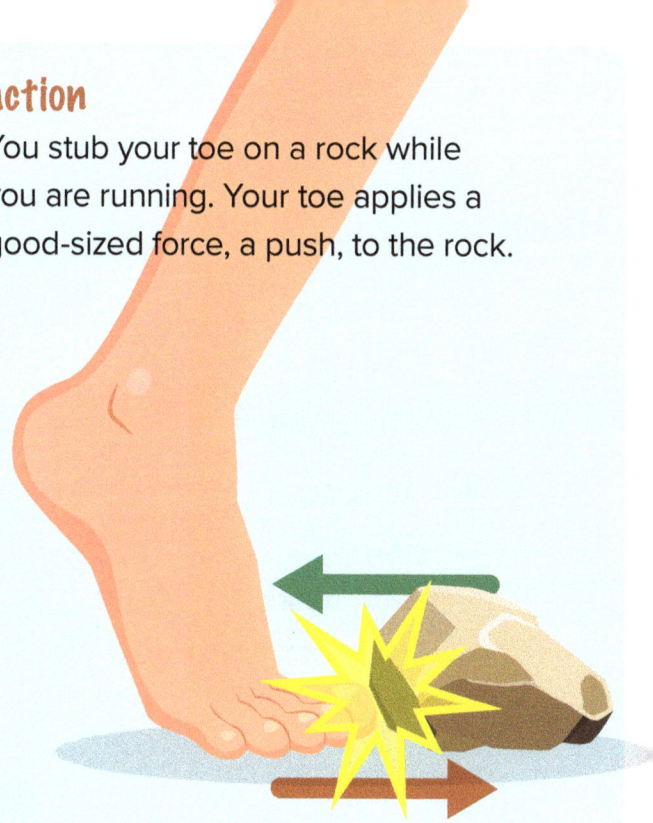

reaction

The rock applies exactly the same amount of force back to your toe, also a push. Ouch!

action

You row your boat across a pond. Your oars push the water.

reaction

The boat moves forward. The harder you row, the further you move forward with each stroke of your oars.

action

You dive into a pool and swim. With each stroke you push water back.

reaction

You move forward!

action

You use force and push down on the diving board.

reaction

The diving board pushes you up into the air.

action

A 5 ball rolls in the direction of a 10 ball. They collide. The moving ball applies a force to the ball that wasn't moving, causing it to bounce off.

reaction

When collided with, the 10 ball causes a push of equal force, so the 5 ball either just stops or bounces off a little bit.

action

You hit a nail with a hammer. Bang! The nail goes into a board.

reaction

At the same time, the nail slows the hammer down and stops it.

If you look around you and carefully observe the motions of objects, big and small, you will start to see Newton's three Laws of Motion everywhere. The truth is, they're happening all over the place, all the time!

7 CHAPTER GRAVITY

One of the most obvious of all our natural laws is the law of gravity. You drop something and it falls. No matter where on the planet you go, no matter what kind of object, if there is no force holding it up, it will fall toward the earth.

But gravity might be even more interesting than you think!

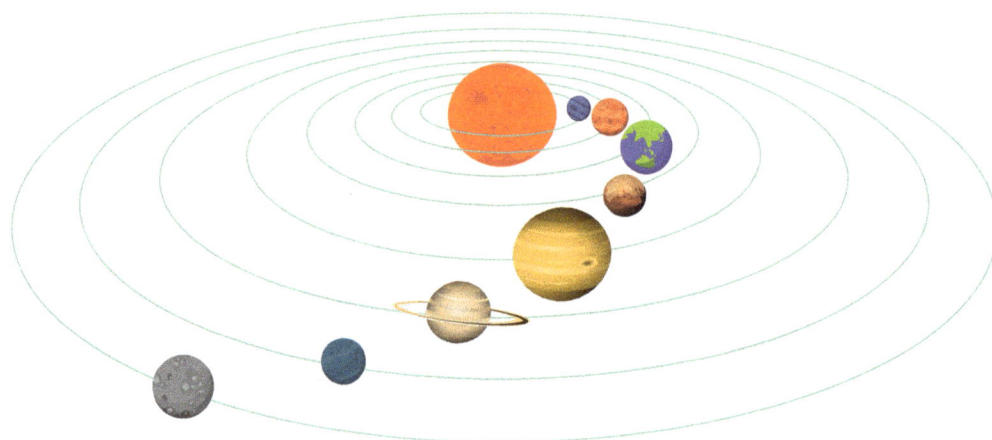

Did you know that *all matter* pulls other matter toward it? It's not *just* the earth or the sun or the moon or some planet like Mars that does this. It's every object that has mass.

A rock pulls other matter toward it. Trees, icebergs, meteors, stars and suns all pull.

At this exact moment, the book in your hands is pulling on every other object in the room, and every object in the room is pulling on it!

What???!!!

If that seems crazy, that's because the pull of your book is so small, you would never notice it, even though scientists can actually measure it.

So even though you can't necessarily see it or feel it, all matter is pulling other matter toward it. In other words, it is attracting all other matter. There is a name for this force, and the name is gravity.

Gravity is the force in the universe that makes all objects attract one another.

You can see this better when one or both of the objects have lots of mass. For example, if you were to drop this book, it would be pulled toward the earth because the earth has so much more mass than the book.

When you have two objects, the gravity force between them depends on the total amount of mass of the two objects.

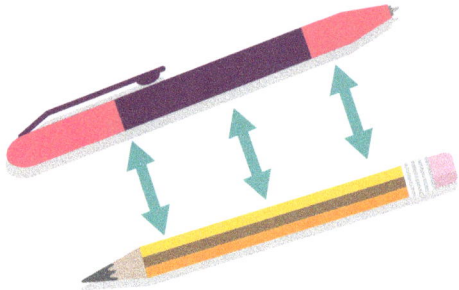

The gravity between small objects, such as pencils, books or rocks, is too small to notice because the total mass is quite small.

Gravity becomes more noticeable as the mass of the objects becomes larger.

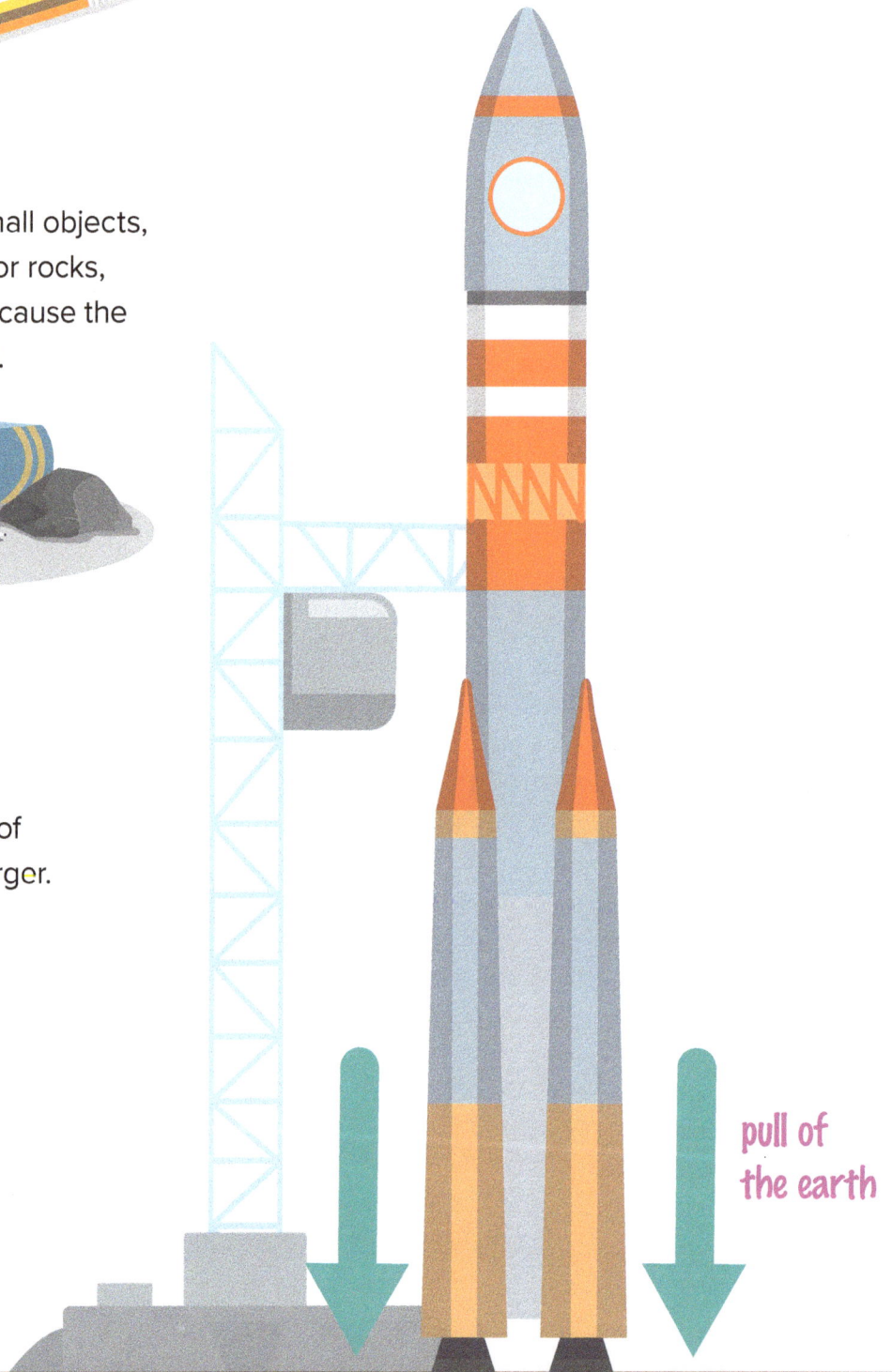

pull of the earth

If you think of the earth as an object, it has a huge amount of mass. So, if you are standing on the earth and let go of a pencil, the gravity will cause the pencil and the earth to move together. We describe this by saying the pencil falls to the earth.

The force of gravity is what holds your body in your chair—the attraction between the matter of your body and the matter of the earth is pretty strong.

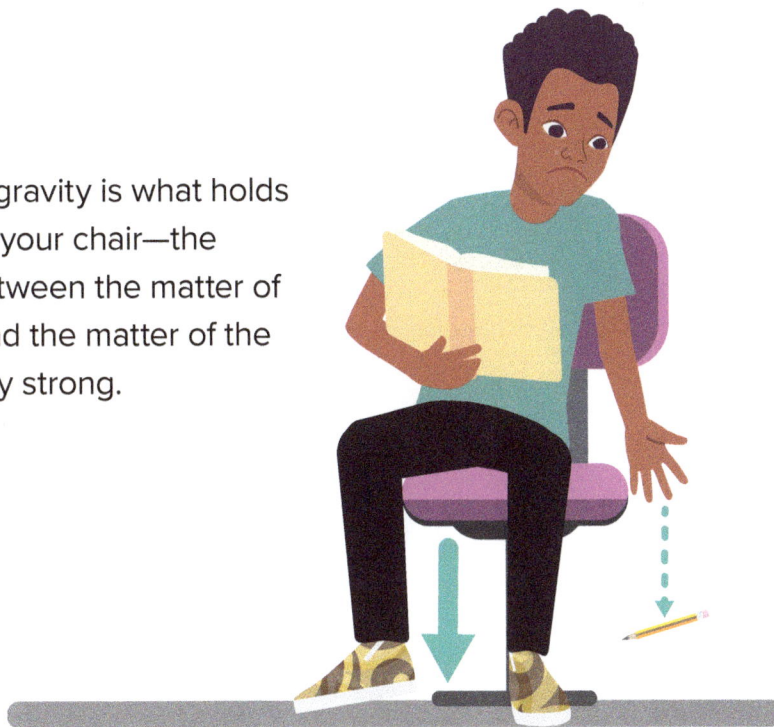

The force of gravity also depends on the distance between two objects. The closer together objects are, the stronger the attraction between them. The farther apart they are, the weaker the attraction between them.

For example, the sun has much more mass than the earth, but we don't feel it pulling on us because it is so far away.

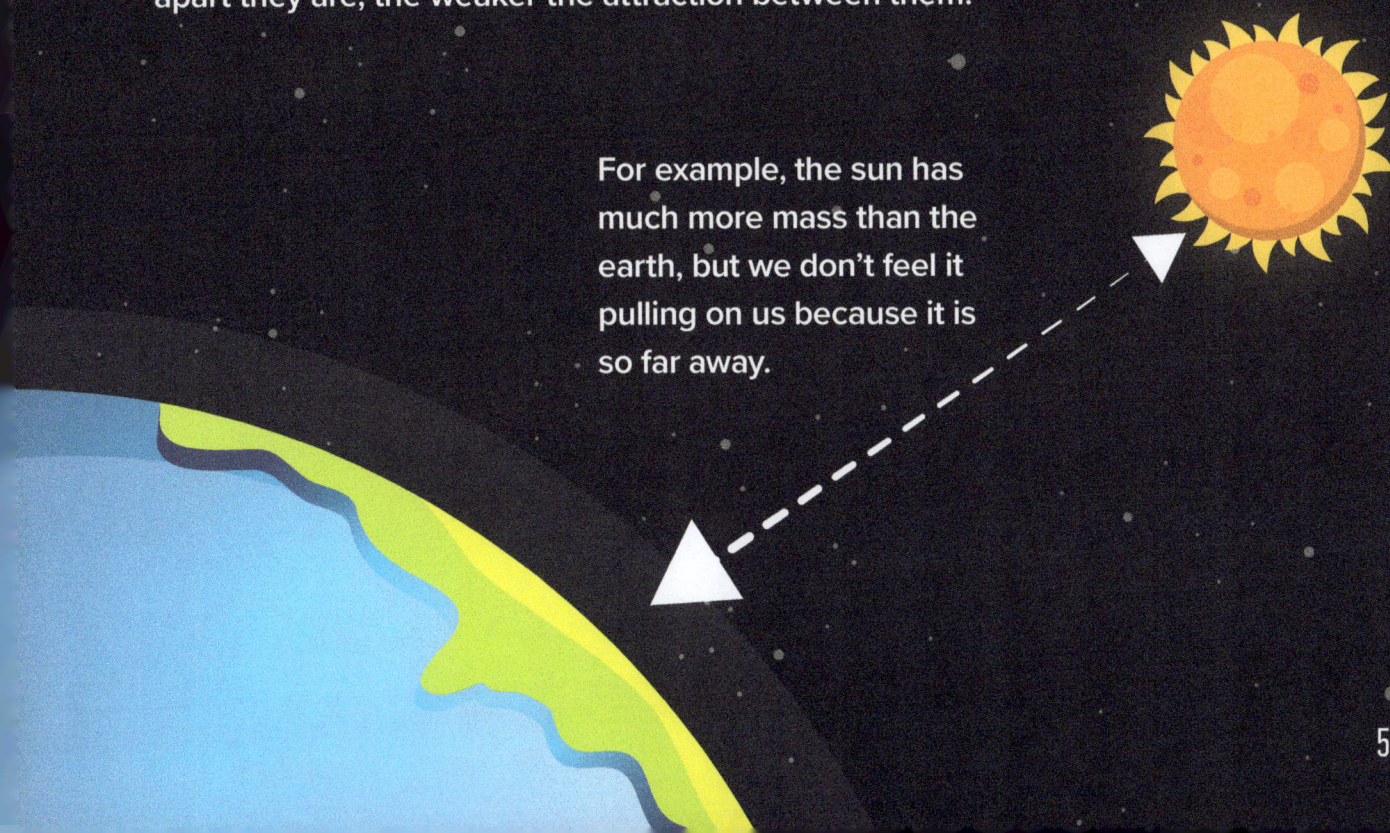

GRAVITY, WEIGHT AND MASS

Weight is a measure of the force of gravity. No matter where a piece of matter is located, it will always have the same number of atoms, the same mass. But the force of gravity acting on that mass can change depending on where it is.

If you asked an astronaut how much she weighed, she might ask, "Where?" Let's say she weighs 140 pounds on Earth. The pull between the mass of her body and the mass of Earth adds up to 140 pounds of force.

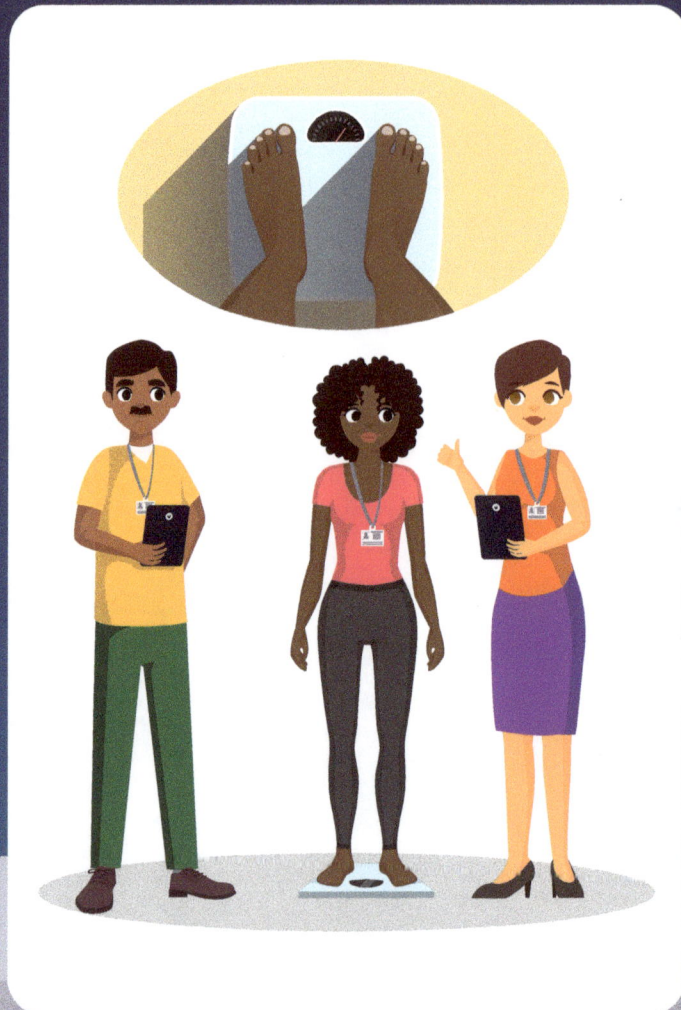

If you put her on the moon and measure the force of gravity there (the pull between the mass of her body and the mass of the moon) her weight would only be about 23 pounds.

The mass of her body remains exactly the same, but the mass of the moon is less than the mass of the earth, so the force of gravity there is weaker and she weighs less.

Weight measures the force of gravity on the object.

The mass of an object doesn't ever change, but the weight of that object depends on how much gravity is pulling on it.

8 CHAPTER FRICTION

Imagine pushing a good-sized rock across a slippery, ice-covered pond. Most likely it would be fairly easy to slide it from one side to the other.

Now imagine trying to slide that same rock along a sidewalk. You'd probably have to push and pull it most of the way. It wouldn't slide easily but you could probably do it.

Now imagine trying to push that same rock along a gravel road. You'd probably give up before getting it very far.

Think about how rough each of these surfaces is. One is smooth and slippery, one is a bit rough and one is full of bumps. Each one offers some resistance to the moving rock. This resistance is called friction. **Friction** is the resistance between two surfaces that are moving across one another.

Each of the three surfaces offers a different amount of resistance to the moving rock. There is less friction, or more friction.

There is very little friction when you push something across slippery ice. There is more when you push it across a sidewalk and even more with a gravel road.

WHAT CAUSES FRICTION?

The surfaces of objects aren't perfectly smooth, even when it appears that they are. Every surface has ridges and bumps. Some are tiny, some are larger.

When you slide one surface over another, the ridges and bumps on one surface push against the ridges and bumps of the other.

So, you experience resistance to the movement of the two surfaces. Things slow down. The objects are harder to move. In other words, there is friction.

It takes force to overcome this. You have to push one surface, or both, hard enough to overcome the resistance of the two surfaces pushing against one another.

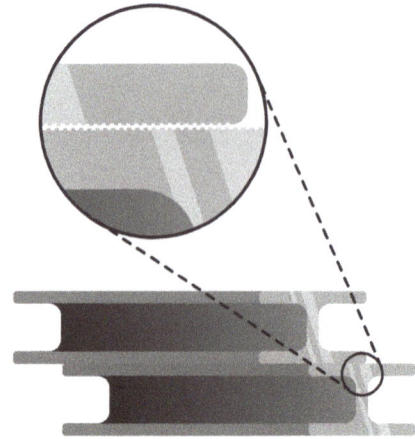

Pushing a rock down the sidewalk is hard because of friction.

Friction can make a job more difficult, but it can also be very helpful. The friction between the floor and the legs of your chair keeps your chair from just sliding across the room.

The friction between your shoes and the ground makes it possible for you to walk without slipping and falling down.

Some surfaces, such as ice, are so smooth there is little friction between them and other objects. This is why it's hard to walk on ice. There is so little friction between your shoes and the ice that it's very slippery.

REDUCING FRICTION

One way of cutting down the friction between two surfaces is putting a layer of slippery liquid between them. This is called **lubrication**. Lubrication doesn't get rid of friction altogether, but it does reduce it.

Lubrication is often used in machines and car engines. Enough oil or grease between engine parts that rub across one another keeps friction from wearing them down. This way they don't have to be continually replaced.

Another way of reducing friction is using wheels or rollers.

Wheels greatly reduce the friction between the suitcase and the floor.

INCREASING FRICTION

Sometimes we want more friction, not less.

For example, we want car tires to grip the road well.

If you are walking down a hill, you want your shoes to grip the path firmly so you don't slip.

The way two surfaces grip together is called **traction**.

If car tires have good traction, they won't slip on the surface of the road even when the road is wet.

A bike with new tires that have good traction will stop quickly when you put on the brakes.

Soccer players wear special shoes with spikes or cleats on the soles.

In some places where winters are cold and snowy, people use special tires called snow tires. These have extra traction on ice and snow.

A bike with worn-out tires that no longer provide good traction will probably skid before it stops.

These increase traction when players run and turn on a soccer field.

HOW FRICTION AFFECTS US

The world would be much different without friction.

If you think about it, everything would be so slippery you wouldn't be able to hang onto it. Imagine what it would be like if pencils and spoons and floors were all as slippery as wet soap!

Friction helps slow objects down.

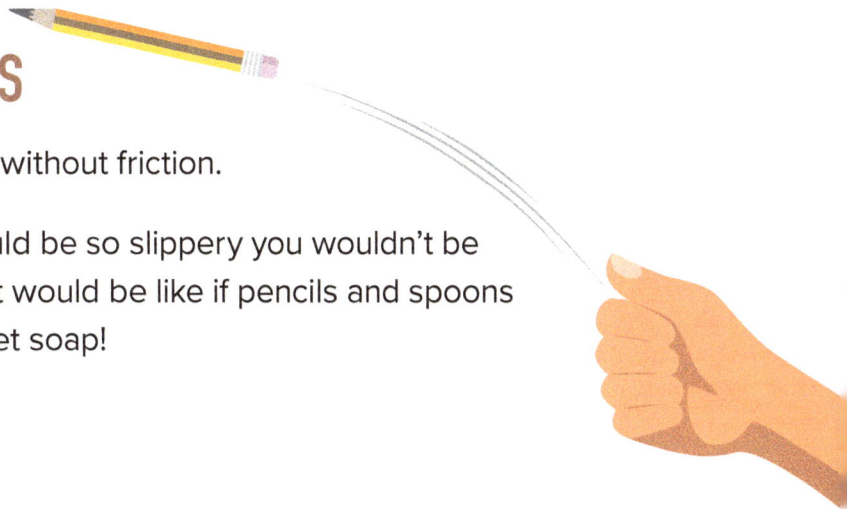

For example, because of friction between a car's tires and the road, it will slow down if the engine stops.

Friction helps us keep things where we want them. It keeps cars on the road, for example. Wearing boots to walk on ice helps keep you from falling.

Taping the handle of a baseball bat keeps it from flying out of your hands.

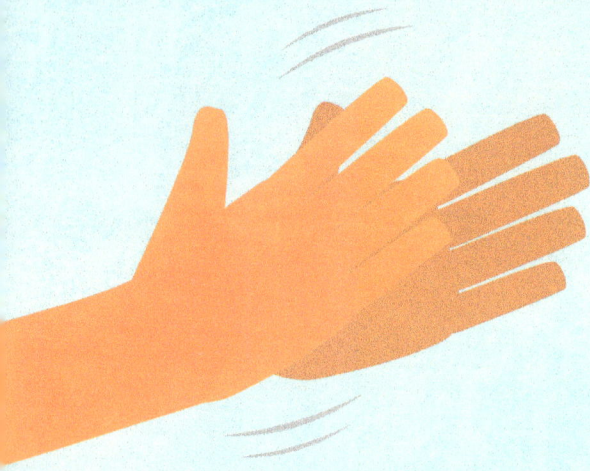

Friction creates heat.
Rubbing your hands
together warms them up.

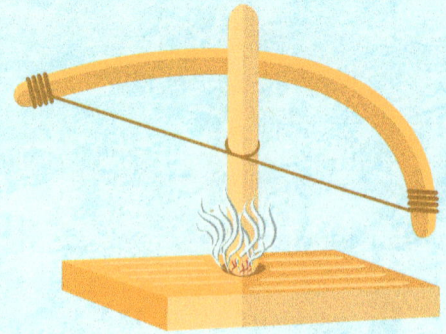

Rubbing pieces of wood
together can create a spark
to start a fire.

Friction makes it possible
to walk and run. The friction
between your shoe and the
ground helps push you along
each time you take a step.

The friction between a bicycle
wheel and the ground helps
the bicycle speed up or slow
down or turn when you want
it to.

When friction gives us trouble, there are three simple ways we can make motion easier.

Make one or both surfaces smoother. Sanding a rough floor with sandpaper, for example, could make it much smoother.

Add lubrication, like oil, grease or soap between the surfaces to make them more slippery.

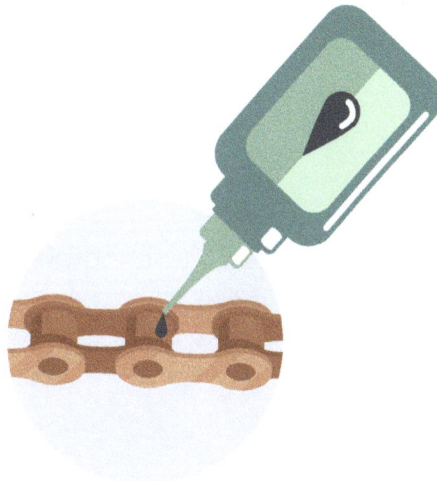

Use wheels or rollers to help an object move more easily.

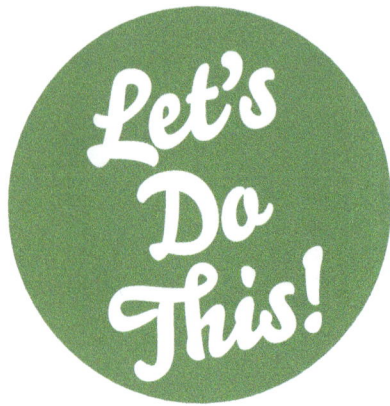

FRICTION AND SURFACES

For this activity you will need

- spring scale that will measure grams and newtons

- small box

- small cart or platform with wheels

- 500 gram weight

- several different flat surfaces. For example, a carpeted floor, a smooth table top, a sheet of metal, a cement floor

- your science journal

Steps

1. Put the 500 gram weight in the box and attach the box to the spring scale.

2. In your journal, note down measurements after each of the following steps.

3. Lay the weighted box and spring scale on a smooth surface. Use the spring scale to pull on the box until it begins to move. Notice how many newtons it takes to get the box moving.

4 Repeat this activity on different surfaces. What do you observe?

5 Put the weighted box on a cart with wheels and use the spring scale to measure the pull needed to move the box. Repeat this on different surfaces. What do you observe?

6 Write any conclusions you made in your journal.

9 CHAPTER MACHINES & WORK

Now that you know more about force and motion, let's look more closely at what a machine is.

We already know that a machine is a piece of equipment used to make work easier. More specifically, **machines** make work easier by increasing or changing the directions of forces.

Let's say you've built a treehouse high up in a tree, and you want to put a mattress in it so you can sleep there. Hmmm...how to get it there?

Well, you could use a rope and pulley to do this. It would be much easier to lift the mattress by pulling down on a rope than it would be to struggle with carrying it all the way up the tree.

By changing the downward force of your pull on the rope to an upward force on the mattress, the rope and pulley would make your job much easier.

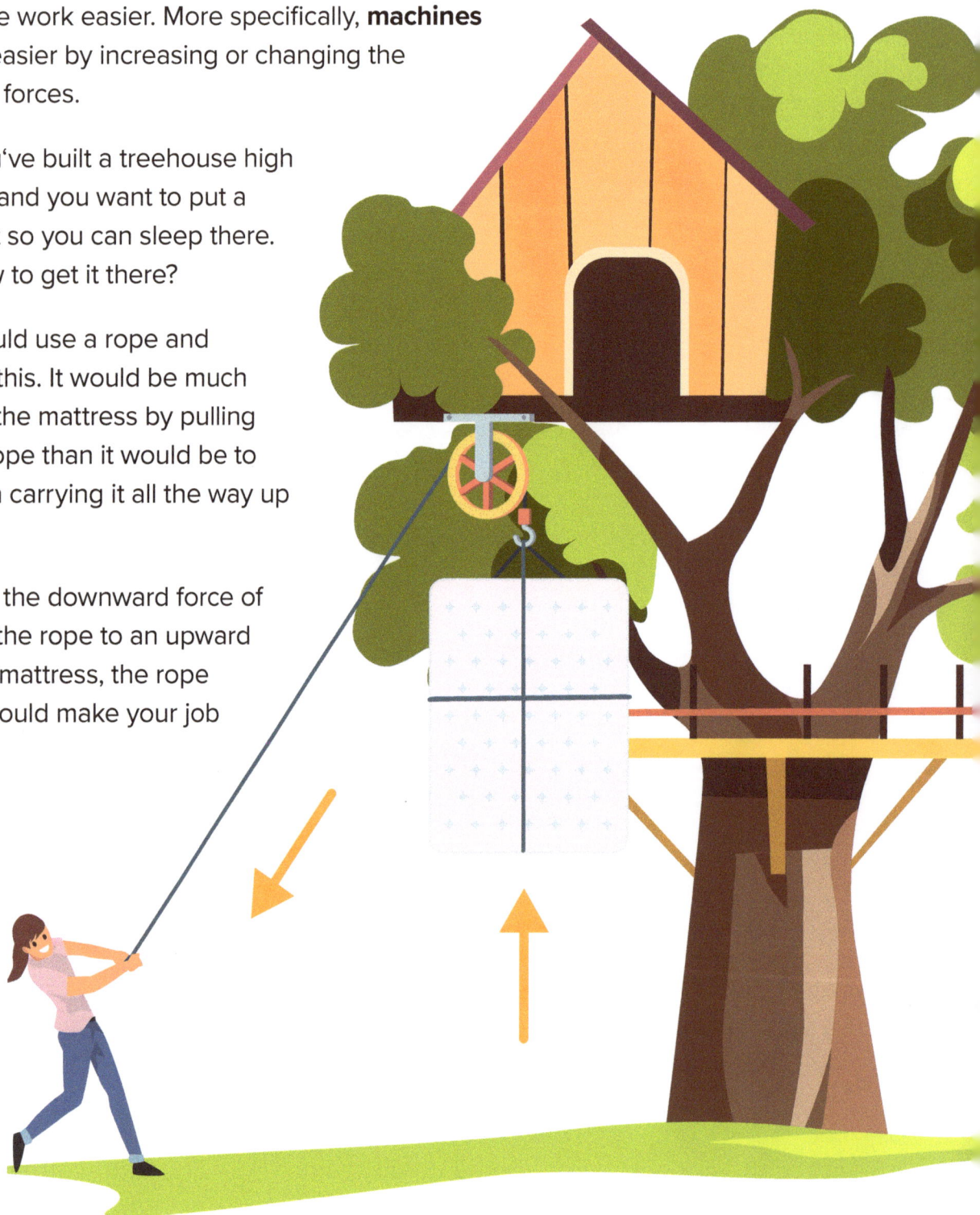

Let's say while digging around outside you find an old chest buried in the ground. Exciting! But how to get it out of the ground to find out what's in it?

You could place one end of a lever under the chest, push down on the other end, and the lever would help push the chest up out of the ground.

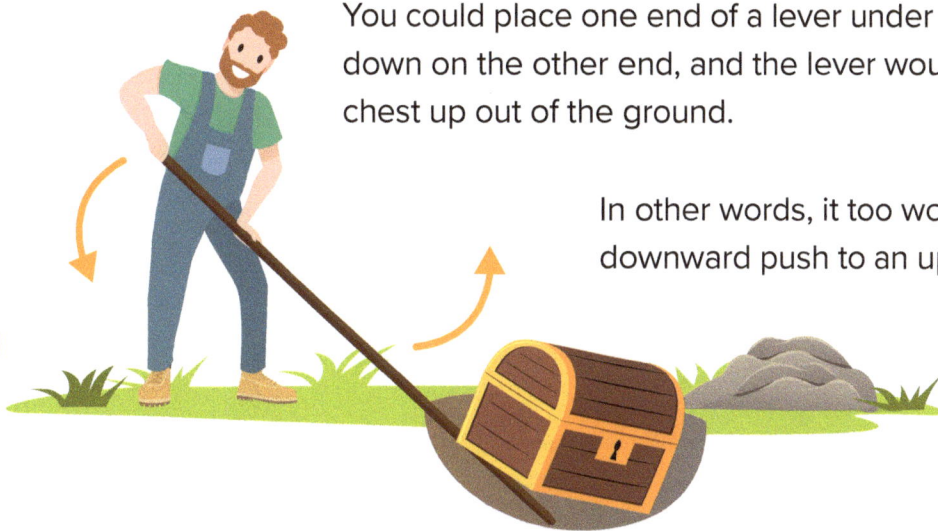

In other words, it too would change your downward push to an upward one.

And now, what do you imagine might be inside the chest. Old coins?

These examples show you what machines do—they increase forces or change the direction of forces to make things move.

WHAT IS WORK?

In science, the word *work* has a very specific meaning. **Work** is moving something. It's using force to make something move across a distance.

You might "work hard" on your studies, but that's a different meaning of work. Here we are talking about a meaning that's very useful in science—using force to move things from one place to another.

When you unload several bags of groceries from your mom's car, you are doing a lot of work.

Why? Because you are using force to move the bags of groceries across a distance, the distance from the car to the kitchen.

WORK IS NOT THE SAME AS FORCE

Simply applying a force is not the same as doing work. Work only happens when a force *moves something* across a distance. If there is no motion, there is no work.

You might push really hard on a heavy wheelbarrow full of dirt to try to get it to move. If you couldn't get it to move, you didn't do any work. Really?!

You might have gotten tired—you may have put a lot of effort into pushing on that heavy wheelbarrow, trying to get it to move. You applied force to it, but even if your force was very strong, it wasn't enough to move the wheelbarrow. According to the scientific definition, you didn't do any work.

If you think about it, a rock that's just sitting on a table is applying some force to the table because gravity is pulling it downward. The weight of the rock on the table is a bit of force. The rock, however, is not moving anything so it's not doing work.

Now imagine tying a rope around the rock, running the rope up through a pulley on the ceiling and back down to a book. If you knock the rock off the table, its weight pulls on the book and lifts it up. The rock did some work. It applied a force to the book and moved it.

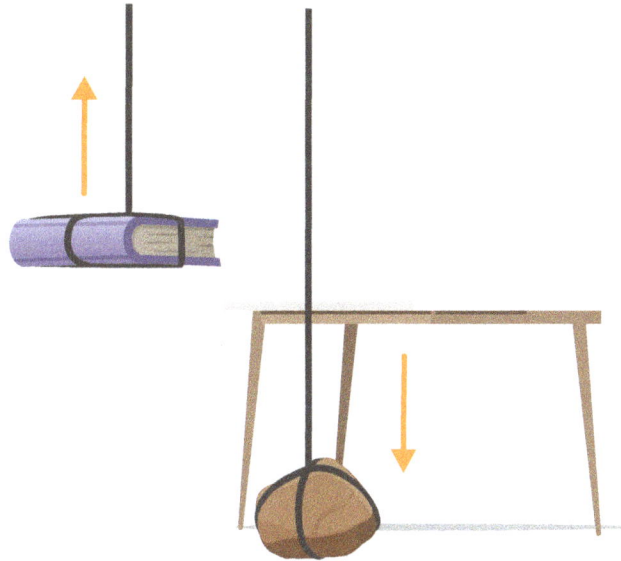

Machines help us do work. They make it easier to move things.

Using a wheelbarrow to move a heavy load is much easier than moving it one shovelful at a time.

Using a plow to dig up a field is a lot more efficient than doing it with a shovel.

Using a hammer to drive a nail into a board is a lot easier than trying to push it in with your fingers.

Riding a bike down a hill is easier and more efficient than walking. And possibly more fun!

67

10 CHAPTER LEVERS

Now that you know more about force and work, let's take a closer look at our first simple machine, the lever.

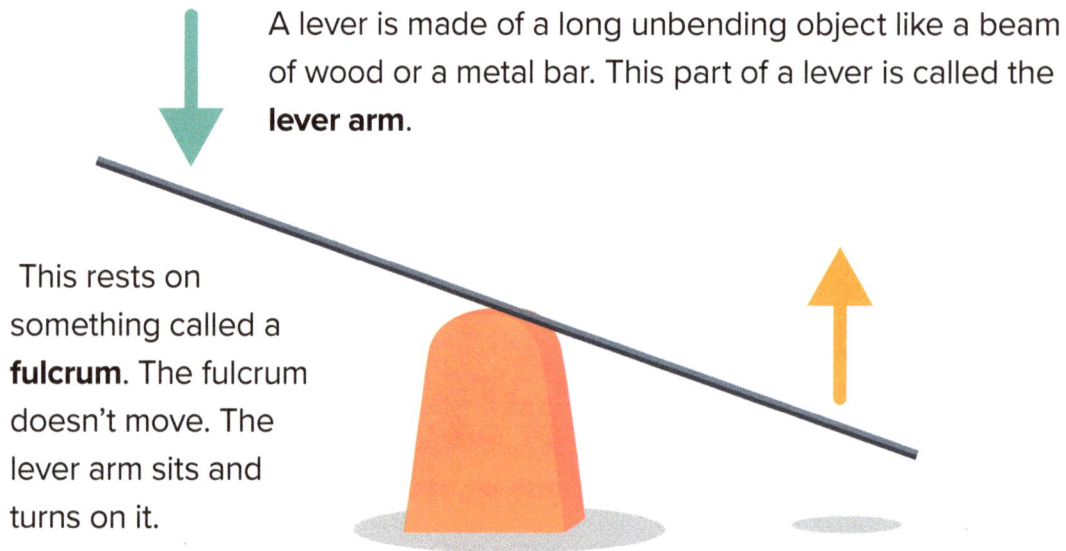

A lever is made of a long unbending object like a beam of wood or a metal bar. This part of a lever is called the **lever arm**.

This rests on something called a **fulcrum**. The fulcrum doesn't move. The lever arm sits and turns on it.

A lever can change the *amount* or the *direction* of a force, making it easier to get a job done. Using a lever you can apply force in a direction that might be difficult otherwise.

As mentioned earlier, a seesaw is a kind of lever. As you know if you've ever played on one, if each person sits in the right place, a small person can lift a heavier person. The lever increases the force of the smaller person.

MECHANICAL ADVANTAGE

An **advantage** is something that increases chances for success or helps make something easier.

A small person has an advantage when trying to crawl through a narrow tunnel. It's easier because they are smaller.

A tall person has an advantage when trying to spot someone in a crowd.

Mechanical means having to do with machines.

If your bicycle has a mechanical problem, that means there is something wrong with its moving parts.

When you get an advantage doing work by using machines, it is called mechanical advantage. **Mechanical advantage** is how much increase in force a machine gives you.

A lever gives you good mechanical advantage. It takes a small force—you trying to lift a rock, for example—and magnifies it into a much stronger force.

This increase in force, the mechanical advantage, is how you can move something heavy like a big rock.

Scientists call what you are trying to move the **load.**

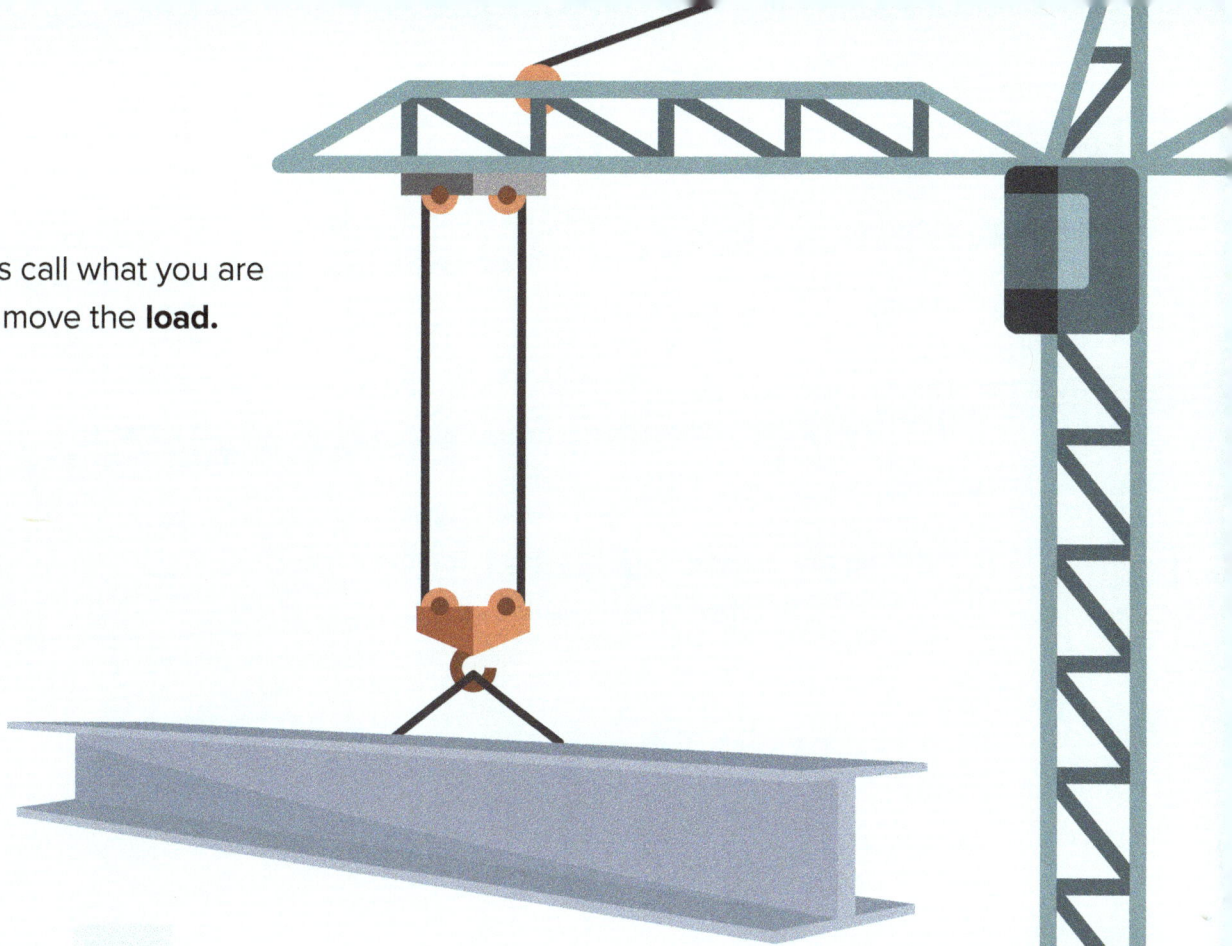

If you were trying to lift a huge beam with a large crane, for example, the beam would be the load.

In the earlier lever example, the rock is the load.

When engineers are using levers, they usually want to get the most mechanical advantage possible.

FULCRUM

What's the secret to getting the best mechanical advantage from a lever?

It's where the fulcrum is placed!

A lever arm has two parts. One goes from the fulcrum to where the force is applied. This is called the **force arm**. The other part goes from the fulcrum to what's being moved, the load. This is called the **load arm**.

The longer the load arm is compared to the force arm, the more mechanical advantage the lever will give you. In other words, placing the fulcrum further from the load will give you more mechanical advantage.

By experimenting with where you place a fulcrum, you will see what works best!

load arm

fulcrum

force arm

With a very long lever and a fulcrum close to the load, you could move an extremely heavy rock. Observing this, an ancient scientist noted that, with a long enough lever and something to use as a fulcrum, he could move the earth.

That would be an amazing mechanical advantage!

USING A LEVER

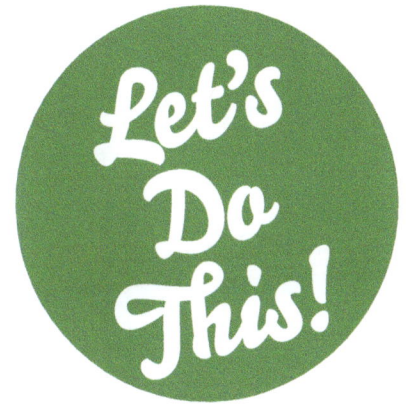

For this activity you will need

- pry bar or strong piece of wood 3–4 feet long

- something to use as a fulcrum

- one or two heavy objects such as a large rock, a log, or a desk. It should be something you definitely could not move without the lever.

- your science journal

- adult supervision

Steps

1 Try to move one of the objects yourself without a lever. Notice the amount of effort you are using.

2 Experiment with the placement of the fulcrum to see what gives your lever the best mechanical advantage.

3 Move the object.

4 Repeat steps 1-3 with another large object, whether it is heavier, lighter or about the same.

5 Write in your journal what you have learned about using a lever and placing the fulcrum.

11 CHAPTER THREE KINDS OF LEVERS

There are three kinds of levers. They work in different ways.

FIRST CLASS LEVER

In a **first class lever**, the fulcrum sits between the force and the load.

Force is applied in one direction and the load moves in the opposite direction. The lever reverses the direction of the applied force. This can make it easier to lift a load.

force

load

fulcrum

SECOND CLASS LEVER

In a **second class lever**, the load sits between the force and the fulcrum. The load and the force move in the same direction. This kind of a lever can also make it easier to lift a load.

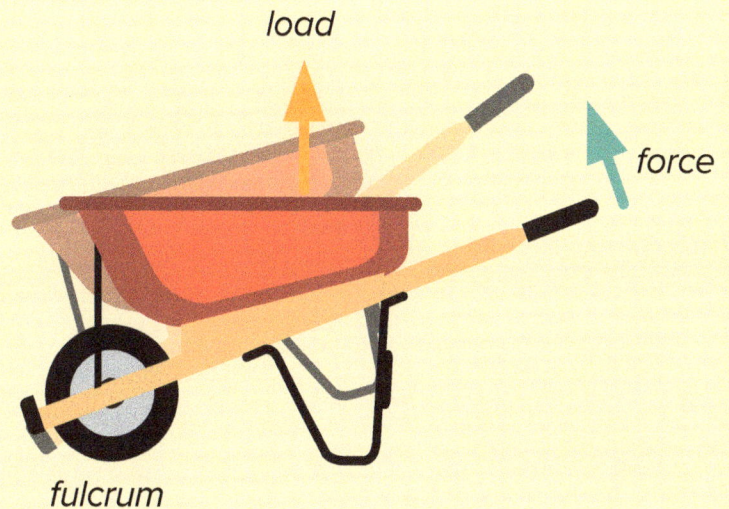

load

force

fulcrum

THIRD CLASS LEVER

In a **third class lever** the force is between the fulcrum and the load. The force and the load also move in the same direction. The load moves further than the force.

A shovel is a third class lever. A person shoveling soil holds the shovel with one hand near the end of the handle and the other partway down.

This is the kind of lever we use most of all. Even your forearm is a third class lever! You use it every time you lift or throw something.

fulcrum

force

load

force

load

fulcrum

The hand at the end of the handle acts as the fulcrum. The other hand acts as the force, and the soil is the load.

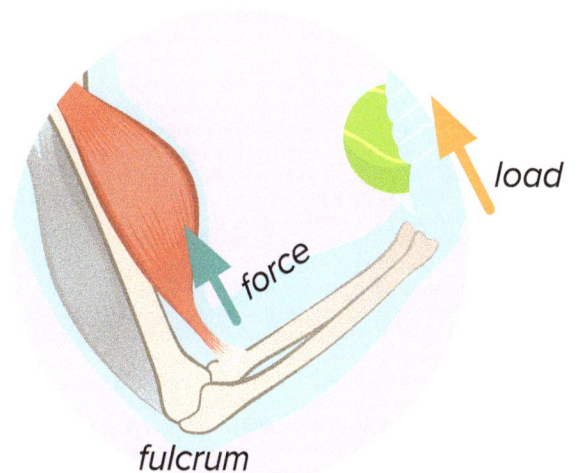

Your elbow joint is the fulcrum. The force is applied where the muscle attaches to the bone. The ball is the load.

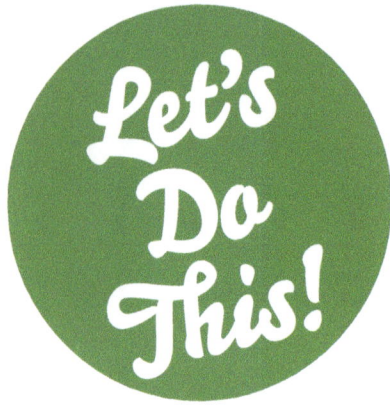

FIRST CLASS LEVER

For this activity you will need

- meter stick or yard stick

- three 1 ¼ inch binder clips

- one foot long piece of ¼ inch diameter dowel

- several books, enough to create two stacks each one foot tall. You could also use two chairs.

- 500 gram weight

- spring scale that measures up to 1,000 grams and 10 newtons.

- your science journal

Steps

1 Use the spring scale to see how many newtons it takes to lift the 500 gram weight. (The 500 gram weight will be the load). Note this in your journal.

2 Set up a first class lever as shown on opposite page. Make sure the middle clip and dowel are right in the center of the stick.

3 Pull down on the scale to bring the lever arm to a level position. Notice how far each end of the lever moves when you do this.

4 Read what the scale says. How many newtons? Note this in your journal.

5 Move the fulcrum to another point on the lever and repeat the steps. Observe what changes. Do this with the fulcrum near the load end (the weight), near the applied force end (the spring scale), and any other positions you want to try.

6 Think about what you did and what you observed. In your journal, explain how changing the position of the fulcrum changed the mechanical advantage. Include sketches if you want to.

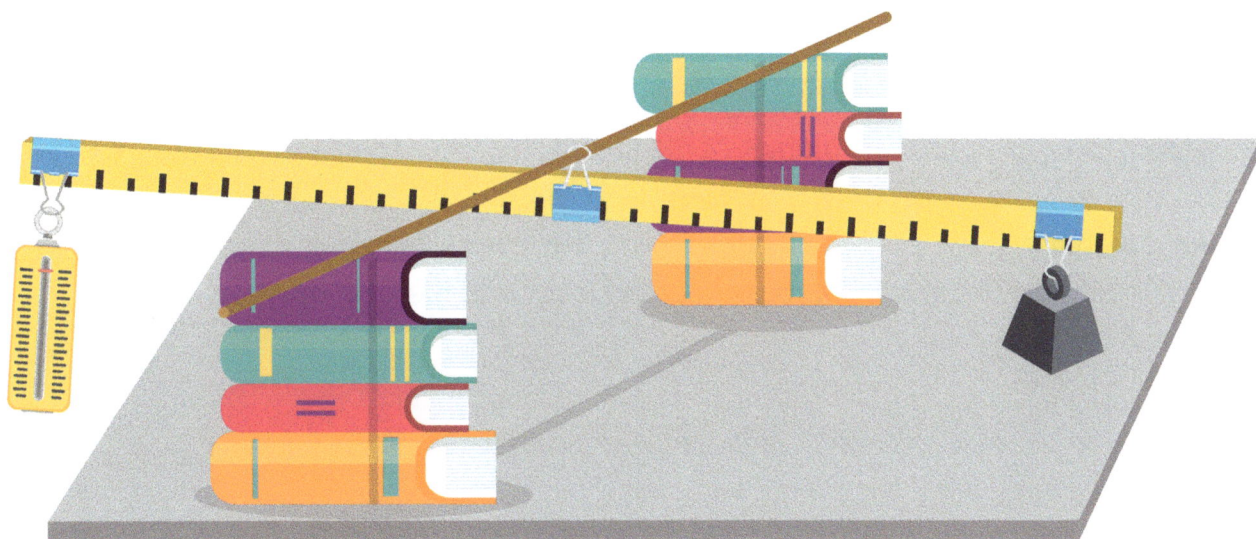

SECOND CLASS LEVER

For this activity you will need

- same as last activity, but this time a 1,000 gram weight

Steps

1. Use the spring scale to see how many newtons it takes to lift the 1,000 gram weight.

2. Set up a second class lever as shown on opposite page. Make sure the weight is hanging from the center of the stick.

3. Pull up on the scale to lift the load, bringing the lever arm to a level position. Notice how far the applied force end of the lever and the load move when you do this.

4. Read what the scale says and note it in your journal.

5. Move the load to another position on the lever and do the steps again. Do this for several different positions.

6. Think about what you did and what you observed. In your journal, explain how changing the position of the load changed the mechanical advantage.

THIRD CLASS LEVER

For this activity you will need

- same as last activity, but this time a 500 gram weight

Steps

1. Set up a third class lever as shown on opposite page. Make sure the scale is positioned in the center of the stick.

2. Pull up on the scale, bringing the lever arm to a level position. Notice how far the applied force point and the load point move when you do this.

3. Read what the scale says and note it in your journal.

4. Move the scale to another position on the lever arm and do the steps again. Do this for several different positions. In your journal, make any notes or sketches you want to.

5. Think about what you did and what you observed. In your journal, explain how changing the position of the applied force changed the mechanical advantage.

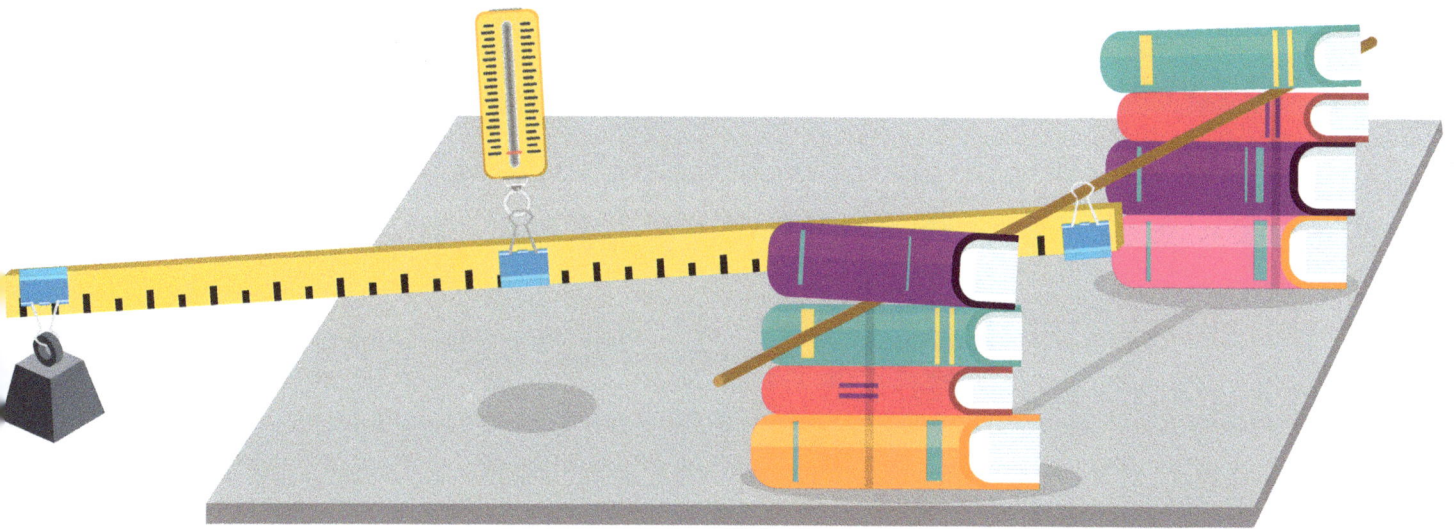

WHEEL AND AXLE

A **wheel and axle** is a circular object, such as a wheel, attached to a rod that is called an axle.

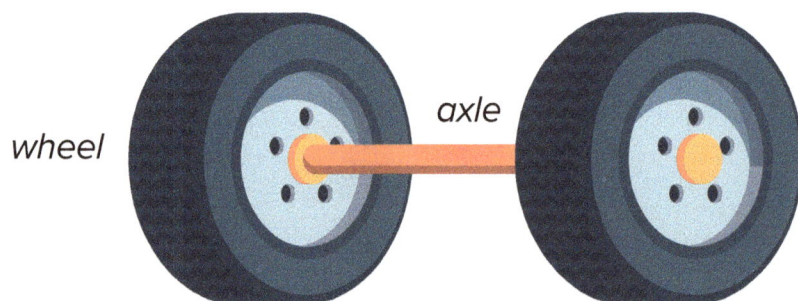

wheel   *axle*

When you turn the wheel, the axle also turns.
Or, if you turn the axle, the wheel turns.

The wheel and axle is a very useful simple machine because it makes it easier to move things. In other words, it provides mechanical advantage.

Here, the wheel is actually a handle that moves around in a circle. The axle is used to raise and lower a bucket. Because the wheel is larger, turning it requires less force and is easier than turning the axle itself.

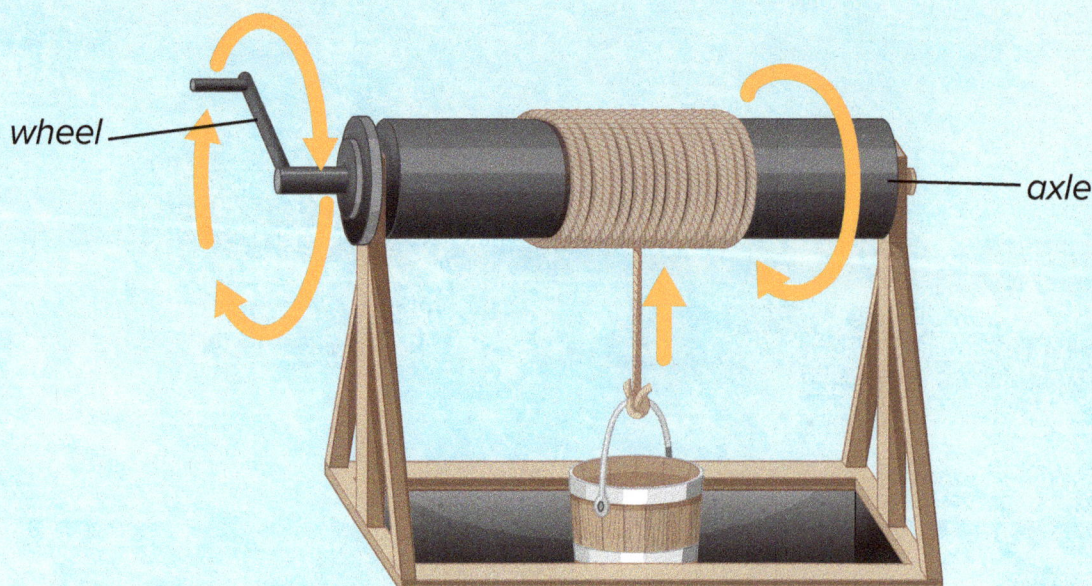

wheel   *axle*

wheel

As another example, it's a lot easier for the wind to turn the blades of a wind turbine than it is to turn the axle they are attached to. The blades, like a propeller on a plane or boat, act as a wheel.

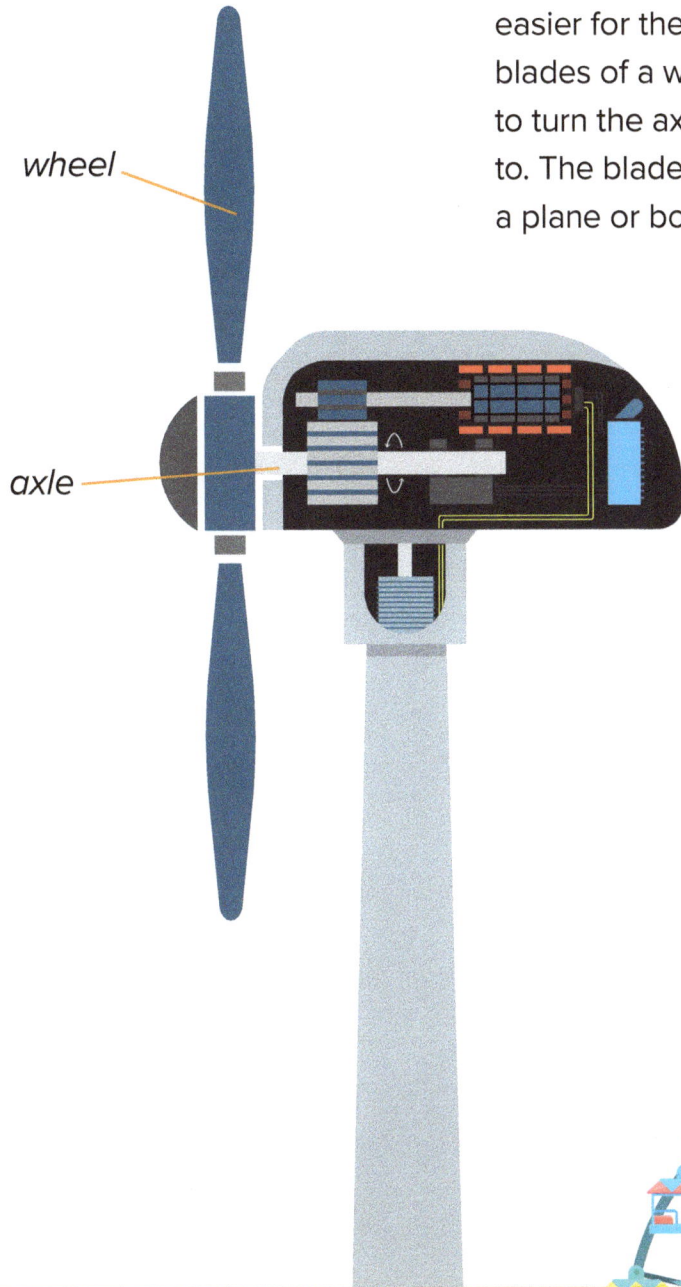

axle

The bigger the wheel compared to the axle, the greater the mechanical advantage.

wheel

axle

Sometimes by turning the smaller axle you turn the larger wheel. Here the force transfers from the axle to the wheel. This is how a ferris wheel works.

WINCHES

A **winch** is a cylinder with a handle used to pull or lift a load. The cylinder is the axle. The handle, also called a **crank**, acts as the wheel. Turning the crank winds a rope or chain around the cylinder, making it easier to lift the load.

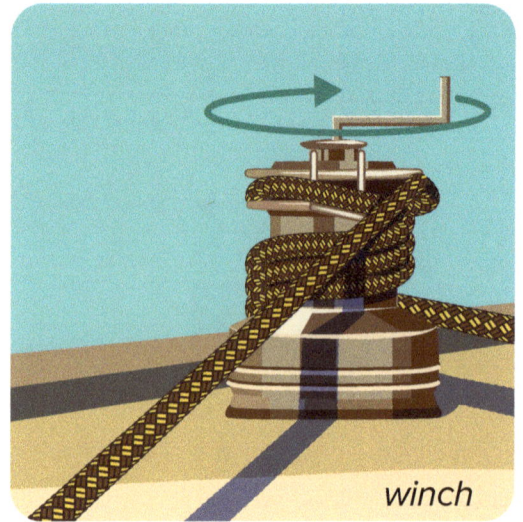

winch

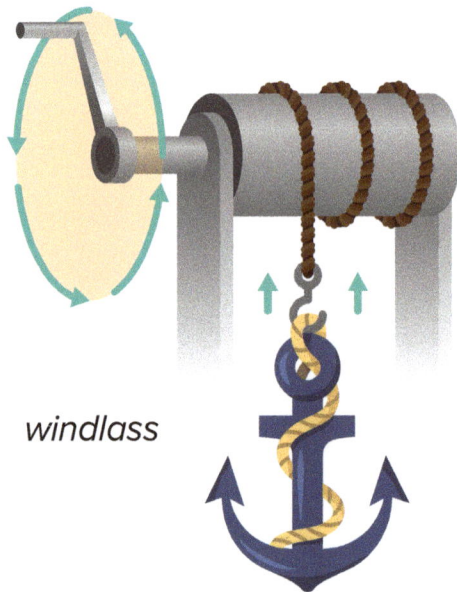

windlass

On boats and ships, a winch used to lower and raise the anchor is often called a **windlass**.

As the crank turns the axle, a rope or cable winds around the axle and this pulls up the anchor.

The longer the crank, the more mechanical advantage you have.

BICYCLES

A bike's front wheel just reduces friction by rolling. It doesn't help you do work.

But the back wheel of a bicycle is a wheel and axle, a simple machine. It's the back wheel that helps you do the work of moving the bicycle.

The pedals on a bicycle turn a special kind of wheel and axle, called a gear. A gear has spikes, or teeth, around its outside edge.

The gear teeth hook into a chain that transfers the force applied to the pedals to another gear on the back wheel of the bicycle. This turns the axle and this is what makes the back wheel move.

OTHER WHEEL AND AXLE EXAMPLES

A doorknob is another example of a wheel and axle. The knob is the wheel, and this is attached to a rod, or axle, inside the door. When you turn the knob, the rod also turns.

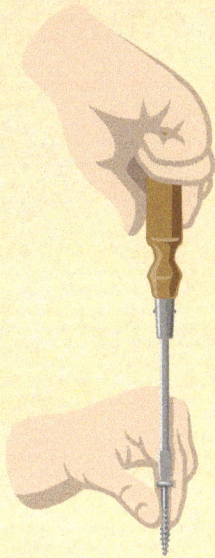

axle

doorknob

Interestingly, a screwdriver is also a wheel and axle. The handle acts as a wheel, turning the shaft, or axle.

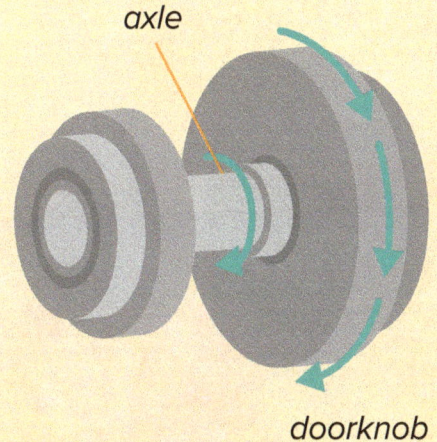

Wherever a rotating motion can be used, a simple wheel and axle machine will give you some mechanical advantage.

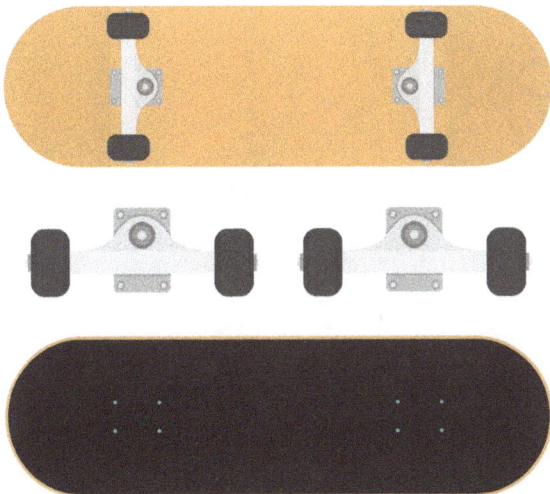

SURPRISE!

Sometimes you will find wheels and axles used in a way that is not this kind of simple machine. For example, if the wheel spins around the axle, like the wheels on a skateboard, then the wheel is simply there to reduce friction. This is useful, but it is not a simple machine because the wheels and axle are not attached. There is no transfer of force between the wheel and axle.

WHEEL AND AXLE

For this activity you will need

- windlass (get this from your teacher)
- 500 gram weight and a 1,000 gram weight
- spring scale that measures up to 1,000 grams and 10 newtons
- your science journal

Steps

Part A

1. Set up the windlass like this:

2. Attach the 1,000 gram weight to the axle cord and the spring scale to the wheel cord.

3. Pull straight down on the wheel cord to lift the load several inches. Notice how far each cord moves. You don't have to measure this exactly.

4. Read what the scale says.

5. Compare the load force and the applied force to get an idea of the mechanical advantage of this set-up.

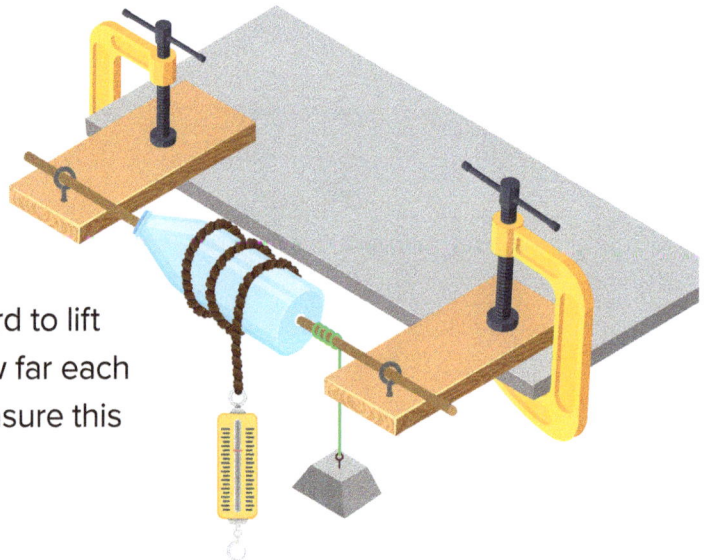

Part B

1 Attach the spring scale to the axle cord and the 500 gram weight to the wheel cord.

2 Pull straight down on the wheel cord to lift the load several inches. Notice how far each cord moves.

3 Read what the scale says.

4 Compare the load force and the applied force to get an idea of the mechanical advantage of this set-up.

5 Sketch or note down your observations. Which set-up gives the most mechanical advantage?

13 CHAPTER PULLEY

Sometimes the simple machine you need in order to get mechanical advantage is a pulley.

There are two types of pulley, the fixed pulley and the moveable pulley.

Remember that a **pulley** is a wheel, hook or ring that you loop a rope around.

FIXED PULLEYS

Fixed means not moving. A **fixed pulley** is attached to something and doesn't move around. It changes the direction of the force.

Let's say you want to lift a heavy weight. The rope goes around the pulley and attaches to the weight. You can lift the weight up by pulling down on the rope.

This doesn't give you any mechanical advantage. But it does allow you to change the direction of a force.

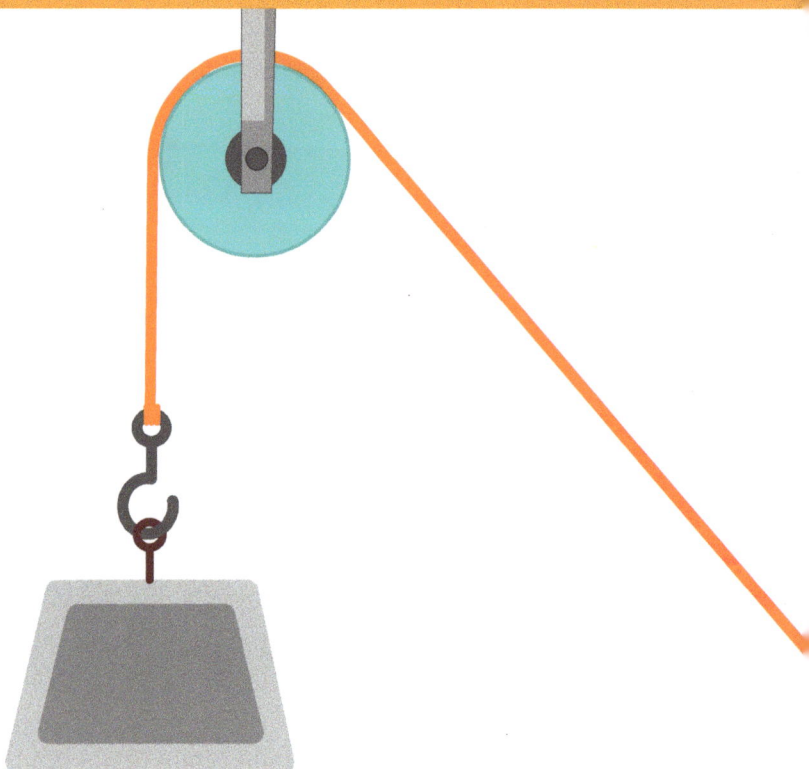

MOVEABLE PULLEYS

The other type is a **moveable pulley**, where a pulley sits on a rope, and the load hangs from the pulley. One end of the rope is attached to something. The force is applied at the free end.

With a moveable pulley arranged like this you get some mechanical advantage. You can raise the load using about half the force it would take without the pulley.

Block and Tackle

A **block and tackle** is a set-up of ropes and one or more pulleys that can be used to lift or pull heavy objects.

In this example of a simple block and tackle, the moveable pulley provides mechanical advantage and the fixed pulley just changes the direction so that you can pull downwards to raise the object up.

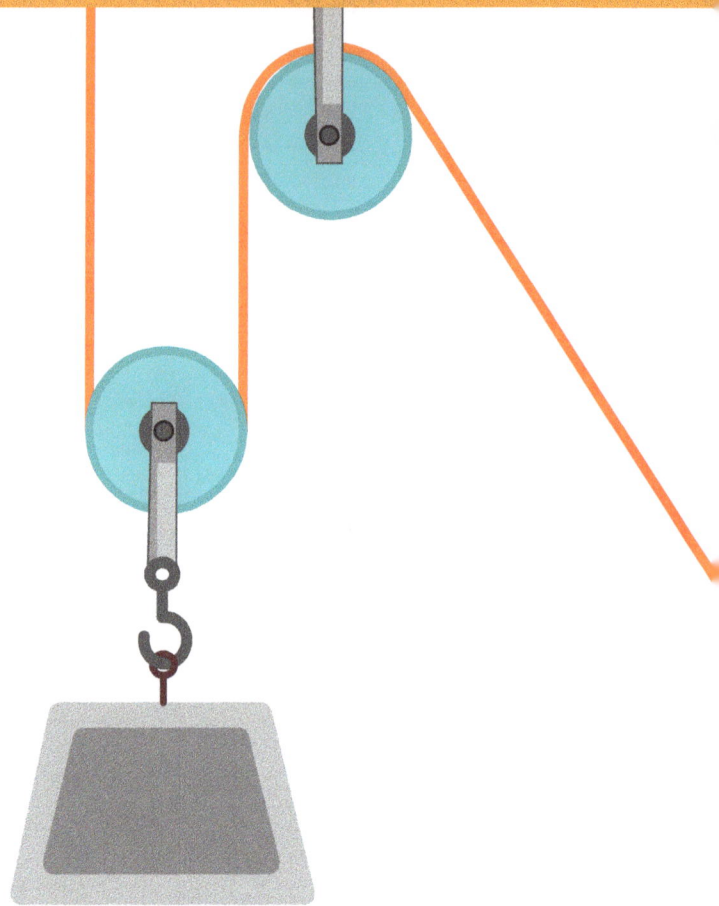

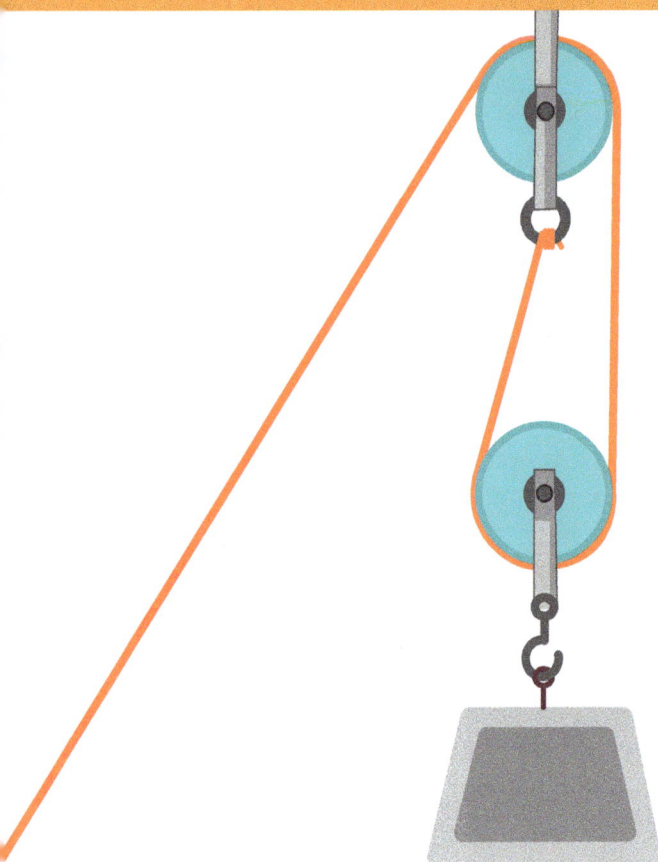

You can also arrange a block and tackle so that pulleys are in line with each other.

To get even greater advantage you can double (or even triple) the number of pulleys. They are usually lined up but don't have to be.

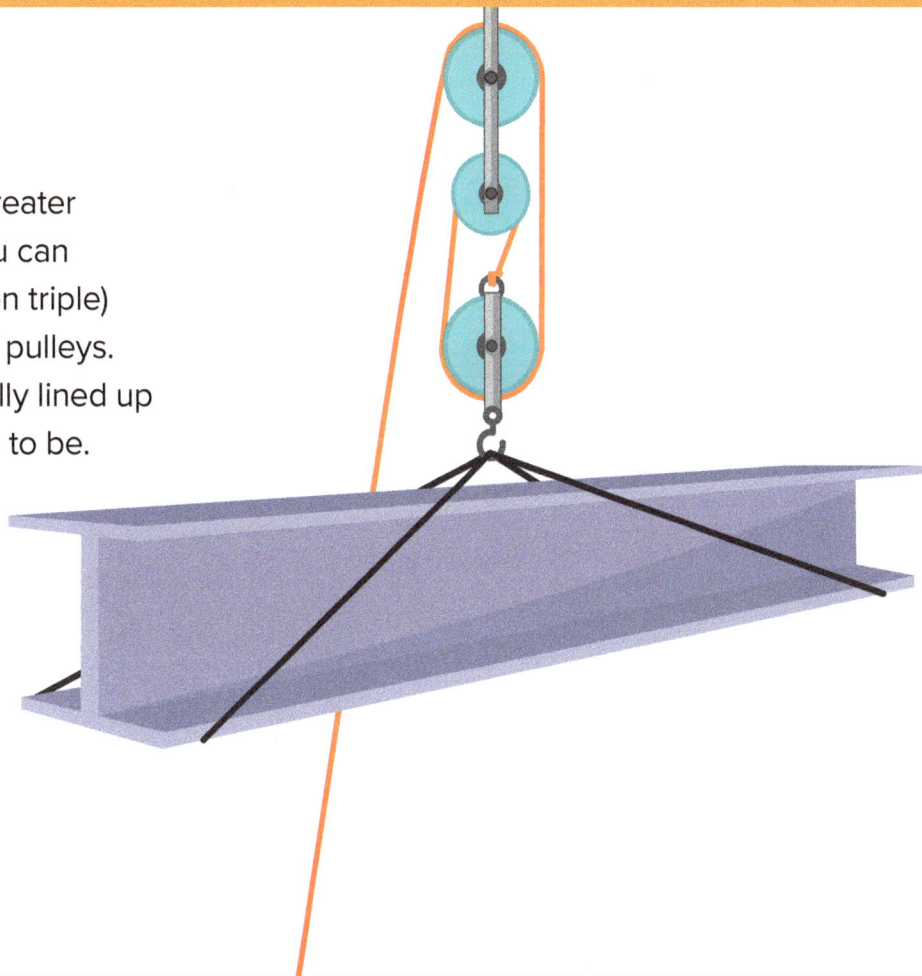

Here's an example using four pulleys! With this kind of set-up, you get a lot of mechanical advantage. You can pull a very large weight without a lot of force. But you will have to pull the rope much farther than the distance the weight will move.

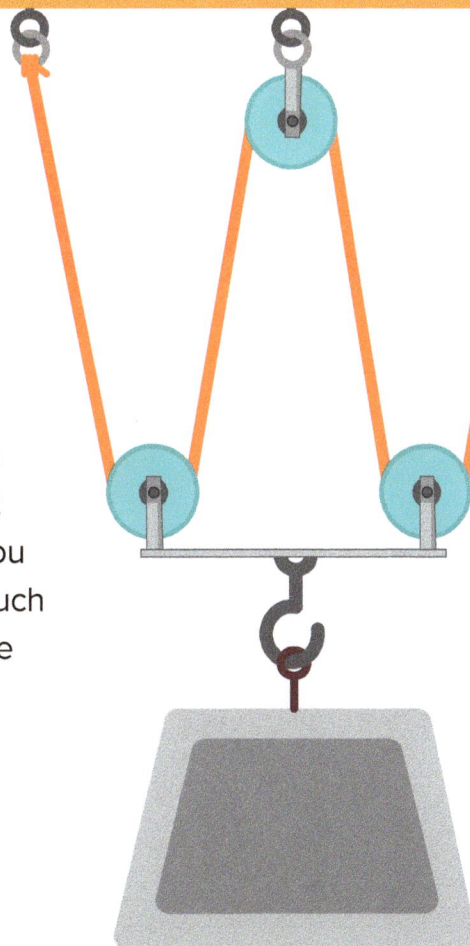

The more advantage you gain by adding pulleys, the more rope you have to pull through the block and tackle. So make sure you have plenty of rope!

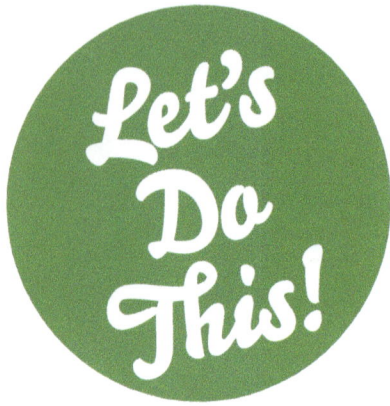

PULLEY

For this activity you will need

- single wheel pulley
- string or cord
- 500 gram weight
- spring scale
- your science journal

Steps

Part A Fixed Pulley

1. Set up the pulley as it is in this picture:

2. Pull on the spring scale to lift the load. Notice how the pulley changes the direction of the force.

3. While holding the load in one position, read what the scale says.

4. Is the advantage you get by using this pulley less force required or change of direction? Note your observations in your journal.

Part B Moveable Pulley

1 Set up the pulley and weight like this:

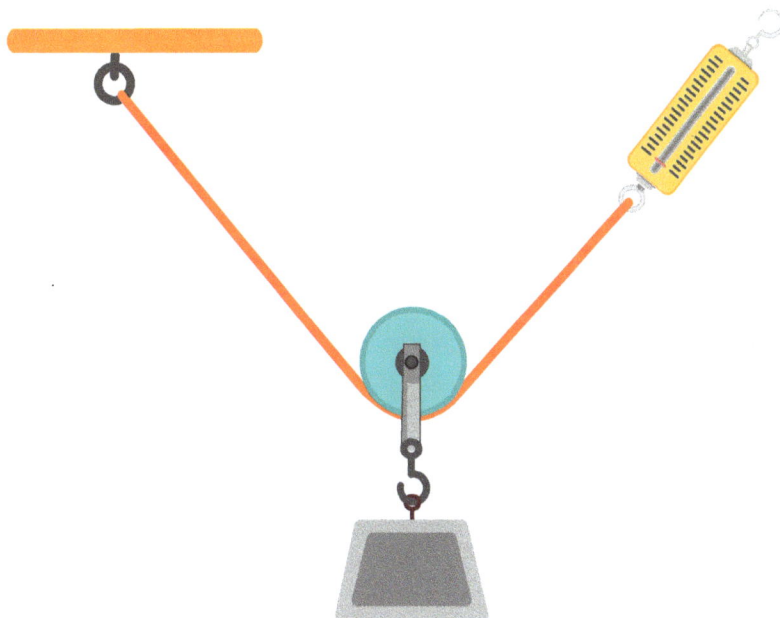

2 Pull on the spring scale to lift the load. Read what the scale says. Notice the distance you pull the cord and the distance the load moves.

3 Is the advantage you get by using this pulley less force required or change of direction, or both? Note your observations in your journal.

4 In your science journal, write down any new thoughts or observations after doing both Part A and Part B of this activity. Feel free to make any sketches you want to show your observations.

BLOCK AND TACKLE

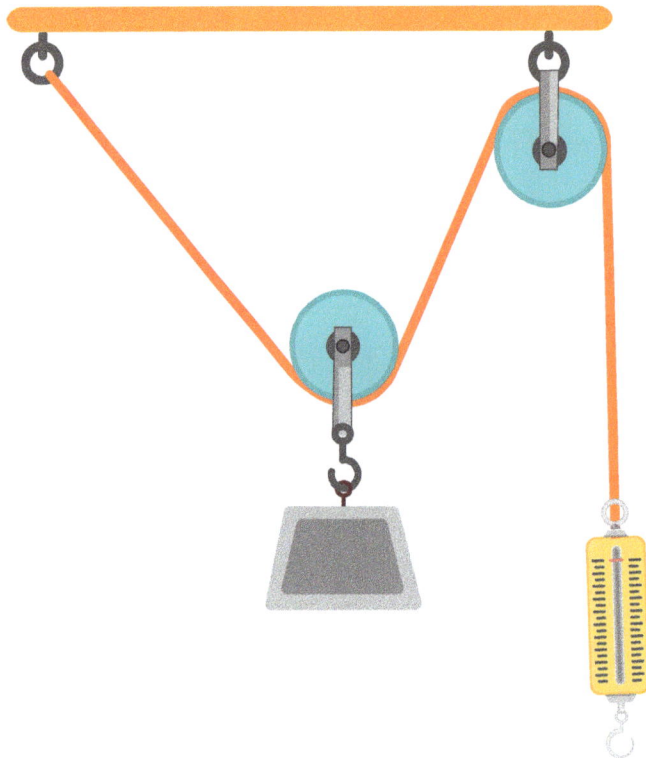

For this activity you will need

- two single wheel pulleys

- two double wheel pulleys

- string or cord

- 500 gram weight

- spring scale

- your science journal

Steps

Part A Single Block and Tackle

1 Set up two single pulleys as they are in the above picture.

2 Pull on the spring scale to lift the load. Notice the distance you pull the cord and the distance the load moves.

3 While holding the load in one position, read what the scale says.

4 Notice that the block and tackle changes the amount of force needed as well as the direction of the force (you pull in one direction and the load moves in another direction).

Part B Double Block and Tackle

1 Set up two double pulleys like this (notice that one end of the cord is tied to the top pulley):

2 Pull on the spring scale to lift the load. Notice the distance you pull the cord and the distance the load moves.

3 While holding the load in one position, read what the scale says.

4 Compare how much cord you pulled through the blocks to how far the load moves.

5 Compare the amount of force needed to lift the load with the fixed pulley and the moveable pulley. Was it more, less, or the same?

6 Compare how far you have to pull the cord and how far the load moves.

7 Sketch or note down your observations in your journal.

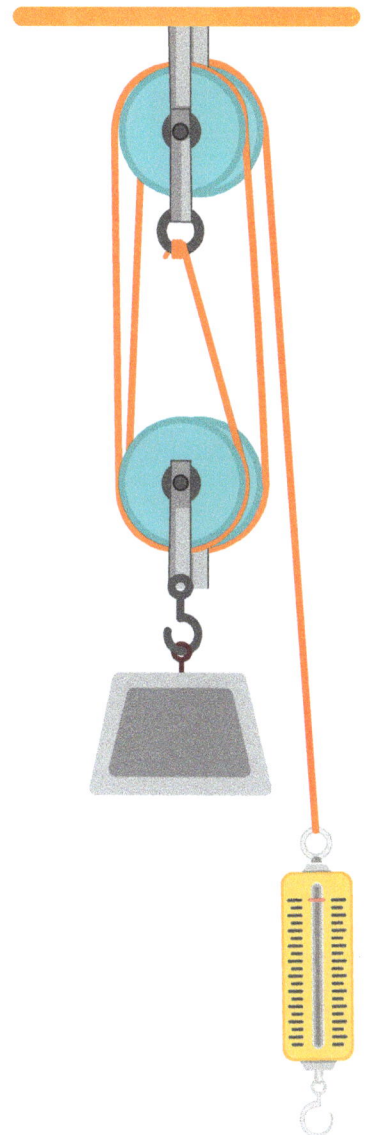

14 CHAPTER INCLINED PLANE

The slanted surface of an inclined plane is a simple machine that can be used to help move things.

A ramp is an example of an inclined plane that makes it easier to move heavy things up or down.

The advantage of an inclined plane is that it takes less force to slide or roll something up or down the ramp than it would to lift it straight up or down against the force of gravity.

A long ramp with an easy slope has more mechanical advantage than a short steep ramp. But you have to move the object a longer distance to get it up or down. It's easier to raise a heavy object up a long, low ramp but you have to push farther.

If you think about it, a ski slope is an inclined plane.

So is a dump truck when it drops off a load.

Even a stepladder is a kind of inclined plane. It helps you move your body from the ground to a higher place.

Let's Do This!

RAMP

For this activity you will need

- 1" thick board about a foot wide and 3-4 feet long

- books for raising one end of the board to different heights

- spring scale

- wheeled cart

- 1,000 gram weight

- your science journal

Steps

1. Hang the weight from the spring scale and notice the number of newtons of force on the scale. Write this down.

2. Attach the scale to the wheeled cart and place the weight on the cart.

3. Find the total newtons needed to pull the cart and the 1,000 gram weight on a flat surface. Write this down.

4. Set the board up so that one end is raised a few inches.

5. Pull on the spring scale to move the cart up the ramp and then hold it in one position.

6. Read what the scale says. Write this down.

7 Repeat the last two steps with the high end of the ramp in different positions (up to a foot or more). Write down your measurements.

8 Notice how the force needed to hold the cart on the ramp changes with different slopes.

9 In your science journal, write down your observations on how an inclined plane gives you mechanical advantage.

15 CHAPTER WEDGE

We know that a wedge is an object that is thick at one end and thinner at the other. At least one of its surfaces is an inclined plane.

A wedge works by changing the direction of an applied force.

For example, an axe is a metal wedge with a long handle. When you use it to chop wood, you take a big swing. The wedge uses the downward force of your swing to cut into a log by pushing aside the wood on both sides of the wedge.

This splits the log in two. The wedge changes the direction of the force.

The mechanical advantage of a wedge depends on how long and how thick or thin it is. A longer, thinner wedge needs a smaller applied force to push it into a piece of wood, but it only makes a narrow cut which may not split the log. A shorter, thicker wedge makes a wider split, but takes more applied force to do that.

It might surprise you to know that the most common wedges in your house are used on food. If you look at the cutting edge of a knife, you'll see that it's actually a wedge!

When you bite into an apple, you use your front teeth as wedges!

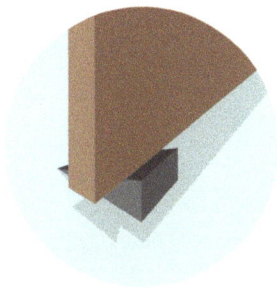

In addition to splitting or opening things, wedges can be used to hold things in place. A doorstop does this.

Wedges are very common simple machines that can help lift things, hold things in place, and change the direction of force in many useful ways.

WEDGE

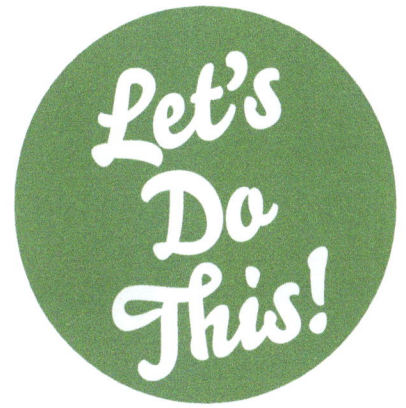

For this activity you will need

- several wooden wedges of different thicknesses
- stack of books at least one foot high

Steps

1. Stack the books against a wall.

2. Insert the tip of the thickest wedge under the stack and push it in so it raises the books. Notice how hard you have to push and how far the books go up.

3. Remove the thick wedge and repeat the last step with a thinner wedge.

4. Do this step with several more wedges and compare the force needed with each one, observing how far the books are raised.

5. In your science journal, write down your observations from this activity.

16 CHAPTER SCREW

A screw is a very narrow inclined plane wound around a central shaft.

When a screw is turned, the turning force and the raised ridges (called threads) push the screw straight into a piece of material like a board or a wall.

A screw changes the force being applied from a rotating force into a straight force.

As usual, to get this change in force, the amount of turning needed is a lot greater than the distance the screw moves into the material.

It takes a lot of motion of your hand and the screwdriver to drive a screw in.

Screws are used to hold things tightly together—the parts of bookshelves, chairs, and tables, for example. These screws change a rotating force into a straight holding force.

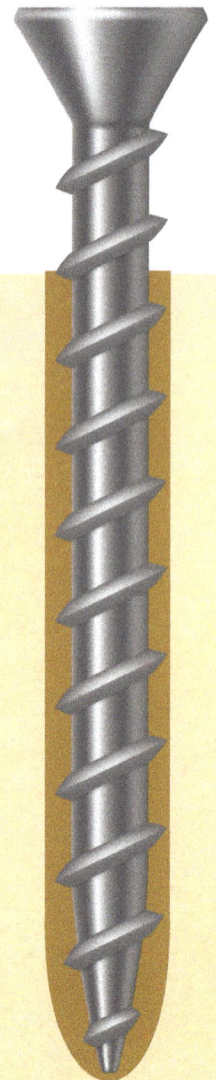

Jar lids and bottle tops are screws. When you turn a jar lid, the turning force on the screw makes the lid go on tight.

The end of a water hose is a screw that holds it tight to an outdoor faucet. If you look at the cap on the gas tank of a car, you'll see it's a screw that holds the cap on tightly.

The base of a light bulb is a screw that holds the bulb in a lamp socket.

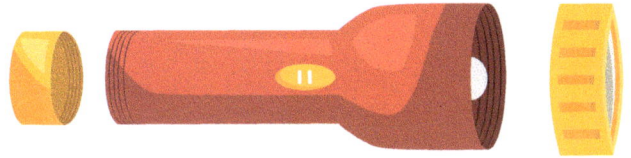

The tops of many flashlights screw on, holding the flashlight together and keeping the batteries in place.

A screw can even be used to move things. If you take a look, you'll see that a swivel chair uses a screw to raise and lower the seat.

A screw can also be used to drill holes in things. Take a look at these—a drill used to dig holes for fence posts, and an electric drill.

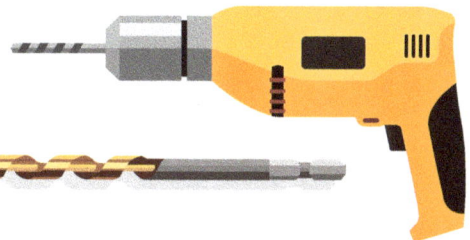

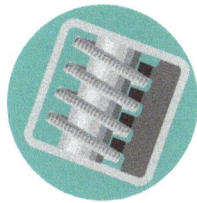

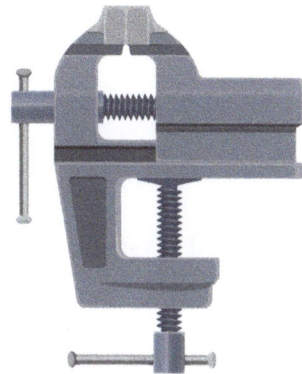

The screw is a useful and very common simple machine.
If you pay attention, you may start seeing it everywhere!

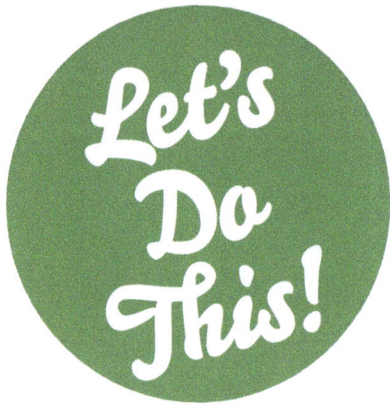

Let's Do This!

Part A

For Part A you will need

- several large wood screws
- several pieces of soft wood
- screwdriver

Steps

1. Look at the screws and find the spiraling inclined plane.

2. Choose one screw and screw it into the wood with a screwdriver. Observe the action of the inclined plane.

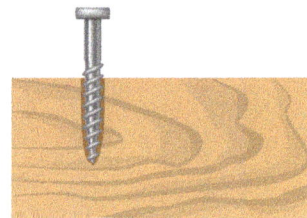

Part B

For Part B you will need

- two pieces of 1-inch thick wood
- C-clamp
- wooden wedge
- hammer or mallet
- your science journal

Steps

1. Examine the C-clamp and experiment with how it works.

2. Have another person hold the two pieces of wood together as tightly as they can.

3. Try to push the two pieces of wood apart by pushing in the wedge. Notice how much force is required to move the pieces of wood apart.

4. Now use the C-clamp to hold the pieces of wood tightly together.

5. Try to push the pieces of wood apart with the hammer and wedge. Can you do it?

6. In your science journal, write down your observations on how a screw can be used to increase force.

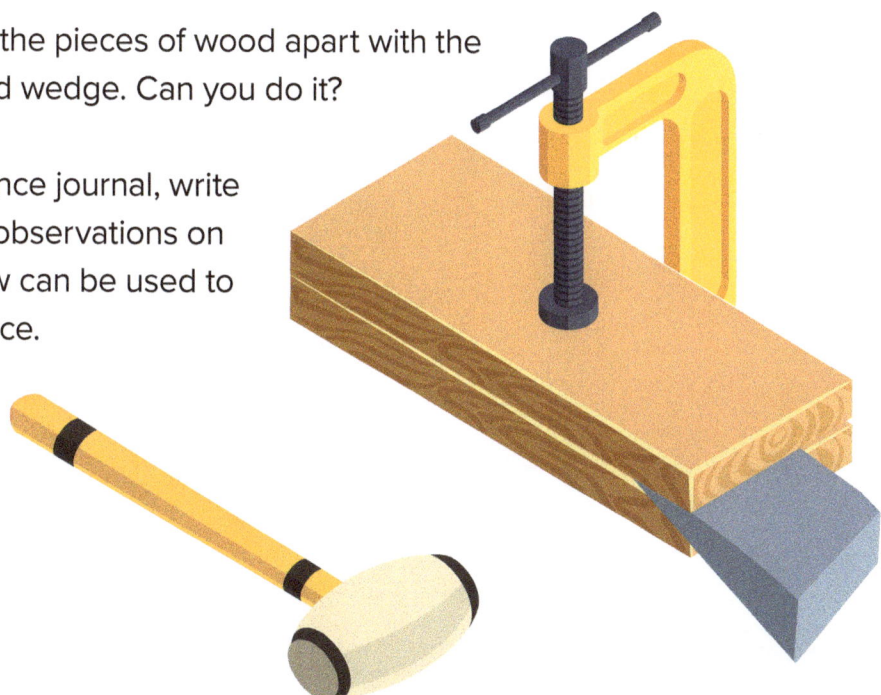

17 CHAPTER COMBINING SIMPLE

You may have noticed that simple machines are often put together into more complex machines that can be used to move things.

Compound means "made from two or more separate things." Engineers call machines that combine two or more simple machines **compound machines.**

A pair of scissors is a compound machine. If you look closely, you will find that it combines two first class levers with wedges along their sides.

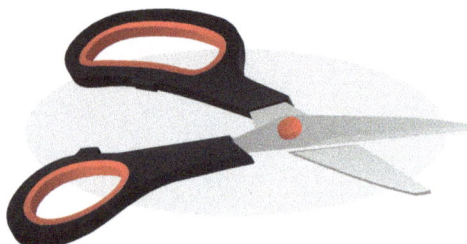

A bulldozer combines a lever, wheel and axle and a wedge. Can you find these?

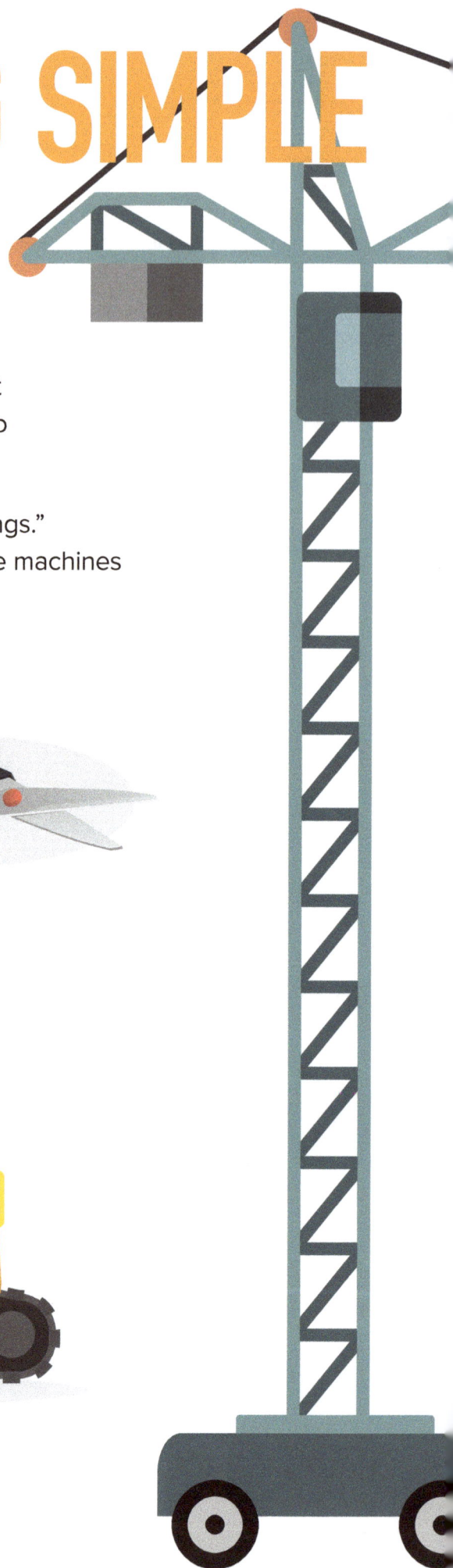

MACHINES

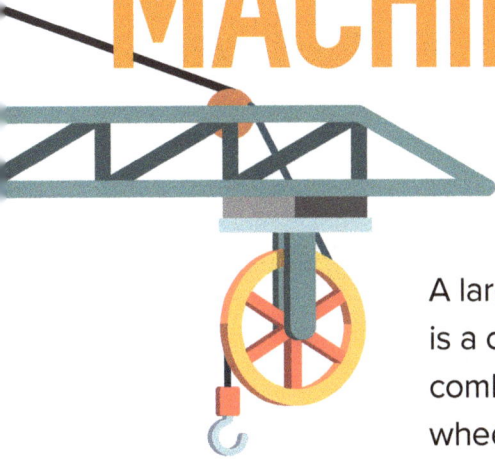

A large construction crane is a compound machine. It combines levers, pulleys and wheel and axle.

Simple machines and compound machines are all around us, helping us do work more easily.

We use them to build houses, bridges and skyscrapers, to plow fields, and to move things from here to there. Any place we want to get a mechanical advantage, we use machines!

Every day engineers are designing new machines to help us do harder and harder jobs more easily. They are using simple machines and compound machines.

What else do simple and compound machines do?

You're a young scientist.
Go find out!

MAKE A COMPOUND MACHINE

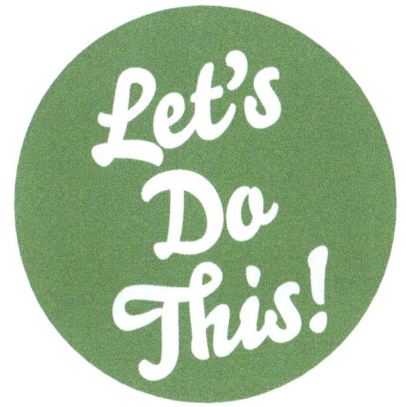

For this activity you will need:

- All the materials already used should be available, including the weights and spring scale.

Steps

1. Combine at least three different types of simple machine to make a compound machine that lets you lift a heavy load more easily.

2. Use weights and your spring scale to see how much your force is increased.

3. Draw a diagram of your compound machine in your science journal. Label all the simple machines it's made of. If you like, give your machine a name.

4. Show off your compound machine to someone!

TEACHER TIPS
How to Construct a Windlass for Student Use

(For use with Wheel and Axle activity, Chapter 12. This can be kept for use with future students.)

Materials

- half-liter plastic bottle (should be thicker plastic that can withstand pressure)
- ¾-inch wooden dowel approximately 2 feet long
- ¼-inch dowel approximately 1½ feet long
- 12-16 ply bakers twine or thin cord, approximately 36 feet
- two pieces of wood approximately 3 inches x 9 inches x 1 inch
- hot glue gun and glue
- drill with ¾-inch bit and ¼-inch bit
- two 1 inch x 3½ inch screw eyes
- two 4-inch C-clamps
- hammer

Steps

1. Drill a ¾ inch diameter hole centered in the bottom of the half-liter bottle.

2. Drill a ¼ inch hole in the dowel 1/3 of the length from one end.

3. Insert the dowel into the bottle, through the bottom and out the top. Position the bottle about slightly off center, closer to one end of the dowel.

4. Secure the bottle to the dowel with hot glue. You may need to fill in with hot glue if the dowel is slightly smaller than the neck of the bottle.

5. Drill or punch a hole in the bottle above the ¼ inch hole in the dowel.

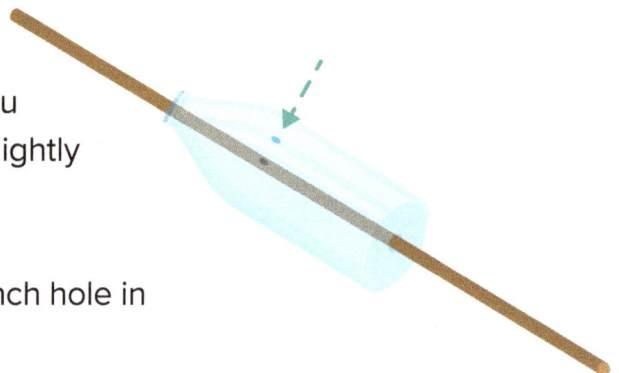

6. Attach a 30-foot section of cord to the center of the bottle with hot glue. Attach a 6-foot section of cord to the dowel. Wind the cords around the bottle and dowel in opposite directions.

7. Screw the screw eyes into the 3 x 9 x 1 inch pieces of wood as shown.

8. Assemble the winch.

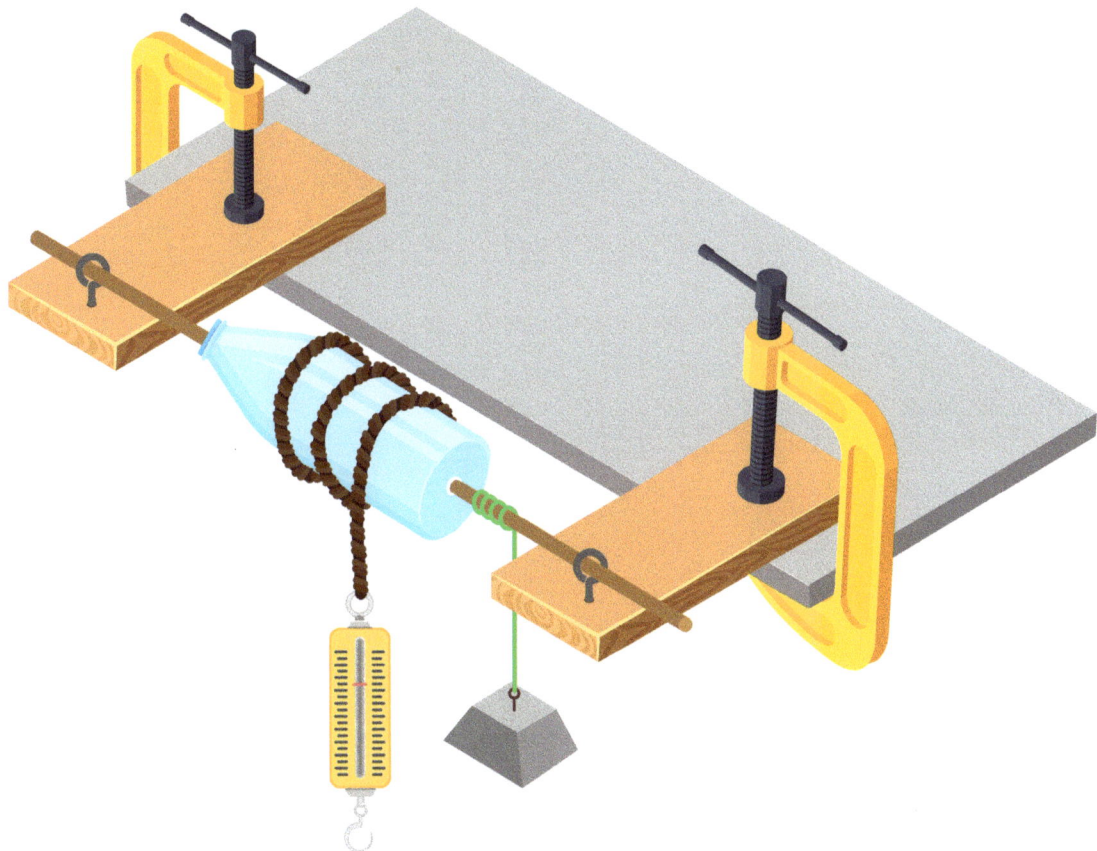

www.ingramcontent.com/pod-product-compliance
Lightning Source LLC
Chambersburg PA
CBHW042033220326
41599CB00045BA/7287